Hydraulic Fracturing Explained: Evaluation, Implementation, and Challenges

Hydraulic Fracturing Explained: Evaluation, Implementation, and Challenges

Editor

Pradeep Sachan

Hydraulic Fracturing Explained: Evaluation, Implementation, and Challenges
Edited by **Pradeep Sachan**

Printed in 2017

ISBN: 978-1-68117-402-0

Library of Congress Control Number: 2015941594

© 2016 by
SCITUS Academics LLC,
616, Corporate Way, Suite 2, 4766,
Valley Cottage, NY 10989

www.scitusacademics.com

This book contains information obtained from highly regarded resources. Copyright for individual articles remains with the authors as indicated. All chapters are distributed under the terms of the Creative Commons Attribution License, which permits unrestricted use, distribution, and reproduction in any medium, provided the original author and source are credited.

Notice

Reasonable efforts have been made to publish reliable data and views articulated in the chapters are those of the individual contributors, and not necessarily those of the editors or publishers. Editors or publishers are not responsible for the accuracy of the information in the published chapters or consequences of their use. The publisher believes no responsibility for any damage or grievance to the persons or property arising out of the use of any materials, instructions, methods or thoughts in the book. The editors and the publisher have attempted to trace the copyright holders of all material reproduced in this publication and apologize to copyright holders if permission has not been obtained. If any copyright holder has not been acknowledged, please write to us so we may rectify.

Contents

Preface vii

Chapter 1   A Continuum Damage Failure Model for Hydraulic
Fracturing of Porous Rocks 1
Amir Shojaei, Arash Dahi Taleghani, and Guoqiang Li

Chapter 2   Numerical Simulation of Complex Fracture Growth
during Tight Reservoir Stimulation by Hydraulic Fracturing 41
Md. Mofazzal Hossain and M.K. Rahman

Chapter 3   Efficient Optimization Framework for Integrated Placement of
Horizontal Wells and Hydraulic Fracture Stages in
Unconventional Gas Reservoirs 81
Xiaodan Ma, Eduardo Gildin, and Tatyana Plaksina,

Chapter 4   A Reinterpretation of Historic Aquifer Tests of Two Hydraulically
Fractured Wells by Application of Inverse Analysis, Derivative
Analysis, and Diagnostic Plots 121
Patrick A. Hammond and Malcolm S. Field

Chapter 5   Evaluation of Methane Yield on Mesophilic-Dry
Anaerobic Digestion of Piggery Manure Mixed with
Chaff for Agricultural Area 173
Dong-Heui Kwak, Mi-Sug Kim, Jae-Seung Kim,
Young-Youl Oh, Soon-Ok Noh, Byung-Ok So,
Su-Young Jung, Su-Jin Jung, and Soo-Wan Chae

Chapter 6   Hydrodynamics of Liquid Film in Helical Tubes 199
Mohammed Salah Hameed and Masab Kadhim Jawad

Chapter 7 The Functional Potential of Microbial
Communities in Hydraulic Fracturing Source Water and
Produced Water from Natural Gas Extraction Characterized by
Metagenomic Sequencing 225

Arvind Murali Mohan, Kyle J. Bibby, Daniel Lipus, Richard W.
Hammack, and Kelvin B. Gregory

Citations 253
Index 257

Preface

Hydraulic Fracturing Explained: Evaluation, Implementation, and Challenges provides updated fundamentals and more recent technology used during a common hydraulic fracturing procedure. Meant for technical and non-technical professionals interested in the subject of hydraulic fracturing, the book provides a clear and simple explanation of the technology and related issues to promote the safe development of petroleum reserves leading to energy independence throughout the world. The developments of hydraulic fracturing technology were coupled to the excitement of the rapid discoveries of gas-shale around the world that could suddenly be produced economically and efficiently. The goal of this book is to advance hydraulic fracturing technology that is effective in its purpose and sustainable in its impacts on communities and environments by bringing together hydraulic fracturing experts not only from the oil and gas industry, but also from other application areas of hydraulic fracturing such as mining and geothermal energy production.

Editor

Chapter 1

A Continuum Damage Failure Model for Hydraulic Fracturing of Porous Rocks

Amir Shojaei[a,b], Arash Dahi Taleghani[b], and Guoqiang Li[a,c]

[a]Department of Mechanical Engineering, Louisiana State University, Baton Rouge, LA 70803, USA

[b]Craft & Hawkins Department of Petroleum Engineering, Louisiana State University, LA 70803, USA

[c]Department of Mechanical Engineering, Southern University, Baton Rouge, LA 70813, USA

ABSTRACT

A continuum damage mechanics (CDM) based constitutive model has been developed to describe elastic, plastic and damage behavior of porous rocks. The pressure sensitive inelastic deformation of porous rocks together with their damage mechanisms are studied for drained and undrained conditions. Fracture mechanics of microcrack and micro-void nucleation and their coalescence are incorporated into the formulation of the CDM models to accurately capture different failure modes of rocks. A fracture mechanics based failure criterion is also incorporated to accurately capture the post fracture crack advances in the case of progressive failures. The performance of the developed elastoplastic and CDM models are compared with the available experimental data and then the models are introduced into a commercial software package through user-defined subroutines. The hydraulic fractures growth in a reservoir rock is then investigated; in which the effect of injection pressure is studied and the simulations are compared with the available solutions in the literature. The developed CDM model outperforms the traditional fracture mechanics approaches by removing stress singularities at the fracture tips and simulation of progressive fractures without any essential need for remeshing. This model would provide a robust tool for modeling hydraulic fracture growth using conventional elements of FEA with a computational cost less than similar computational techniques like cohesive element methods.

GRAPHICAL ABSTRACT

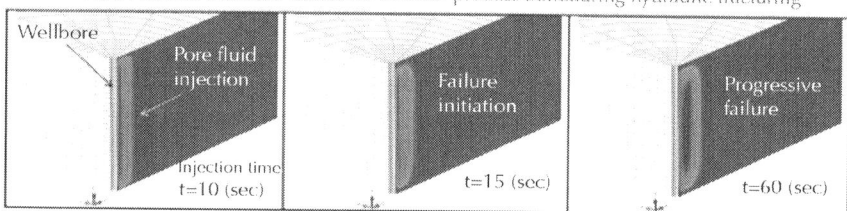

INTRODUCTION

Damage process through micro-cracking and -voiding is identified as one of the main reasons for nonlinear deformation and failure of brittle granular materials like rocks, concrete, and ceramics (Bažant and Planas, 1998 and Voyiadjis et al., 2005). Two major consequences of formation of micro-cracks are materialsoftening (additional compliance) and induced anisotropy due to the directional nature of the damage process. Extensive laboratory and theoretical studies have been conducted in the last half of the century to understand and predict this type of failure in different materials. These approaches range from methods explicitly considering the micro-cracks as single inclusion and non-interacting inclusions (Mura, 1987 and Nemat-Nasser and Hori, 1993), to multiple interacting (Moschovidis and Mura, 1975), to homogenization (Doghri et al., 2010), and micromechanical techniques which consider the self-consistency (Budiansky and O'Connell, 1976, Needleman, 1987, Shojaei and Li, 2013 and Bedayat and Dahi, 2014), or complimentary strain energy method (Sayers and Kachanov, 1991).

In the case of porous rocks, damage-related irreversible deformation may develop due to residual opening of micro-flaws after unloading, which may affect the hydraulic conductivity of the rock. Applying the above-mentioned methods for drained deformation of porous materials should be straightforward; however, there are many examples of failure of geomaterials in conditions different from drained conditions, where coupling between mechanical deformation and fluid pressure changes should be included. The true physics of this coupling is established in the theory of poroelasticity and poroplasticity (Biot, 1956a, Biot, 1956b and Rice and Cleary, 1976). To describe the behavior of anisotropic porous material, (Thompson and Willis, 1991) have extended the Biot's theory to include the anisotropic material and hydraulic properties. This model was later adapted into a poroelastic damage, or more precisely a continuum damage mechanics (CDM) model, by Shao (1998). The main advantage of CDM based porous models is their capability to capture crack initiation, propagation, interaction and possible branching in an integrated framework, which allows material properties evolution during failure.

The mechanical behavior of the earth's shallow crust is complicated with elastic and plastic deformations which are coupled with pore fluid diffusion of infiltrated ground water and hydrocarbons. The porous rocks in this condition are saturated with liquid, i.e. undrained condition. Furthermore, the rock elements are under a compressive stress field from overlaying rocks. Failure and deformation mechanisms of rocks in the drained and undrained conditions have been under intensive research. For example, an asymptotic solution for crack growth in elastic–plastic undrained rock has been developed by Radi et al. (2002). Shao and coworkers have studied the elastoplastic deformation of rocks (Shao and Henry, 1991) and their damage mechanisms (Shao, 1998 and Shen et al., 2013). Viscoplastic creep of rocks around a lined tunnel has been investigated by Cristescu, 1988 and Krajcinovic and Mastilovic, 1999 have developed statistical models of brittle deformation in rocks.

Non-local forms of damage mechanics have been developed to remove the mesh dependency in FEA problems, associated with the damage softening (Marotti de Sciarra, 2012, Voyiadjis and Faghihi, 2012 and Voyiadjis and Mozaffari, 2013). The mesh deletion may result in strain-softening behavior, leading to strain localization, which is called ill-posed constitutive relations (de Borst and Sluys, 1991, Sluys and de Borst, 1994 and Shojaei et al., 2012). This latter phenomenon results in a strong mesh dependency of the FEA in which the dissipated energies decrease upon the mesh refinement steps. This issue is usually alleviated by introducing a characteristic length into the CDM formulation. In FEA packages this characteristic length can be correlated to the element size, and the energy dissipated during the damage process is specified per unit area, not per unit volume (ABAQUS, 2011). Then the damage dissipated energy is treated as an additional material parameter and it is used to compute the displacement at which full material damage occurs. This formulation ensures that the correct amount of energy is dissipated and greatly alleviates the mesh dependency. Consequently the softening response of the constitutive law is expressed based on a stress–displacement relation, in which the displacement is computed from the energy descriptions, e.g. plastic and damage dissipation energies, instead of the ill posed constitutive relations.

Also CDM models for ductile material have been studied extensively in the literature, where various damage mechanisms depending on stress intensity, stress triaxiality and dynamic energy density are incorporated to study dynamic fracture, wear and cyclic loading problems (Malcher et al., 2012, Beheshti and Khonsari, 2011, Kruch and Chaboche, 2011, Aghdam et al., 2012, Brünig et al., 2013 and Shojaei et al., 2013). Progressive failure analysis of polymeric matrix composites within CDM framework has also been a hot topic in recent years (Xu and Li, 2010, Li and Xu, 2011, Voyiadjis et al., 2011, Li and Shojaei, 2012, Naderi et al., 2012, Voyiadjis et al., 2012a, Voyiadjis et al., 2012b, Voyiadjis et al., 2012c, Hansen et al., 2013, Shojaei et al., 2012, Shojaei and Li, 2013 and Kahirdeh and Khonsari, in press). Failure of granular material based on a Discrete Element Method has been investigated by Nicot et al. (2012), and stress and velocity profiles in well-developed dense granular flows has been studied by Kamrin (2010). In the case of Rock Mechanics Khan and co-workers have developed constitutive models to capture elasto-plastic behavior of Berea sandstone under a large range of confining pressure (Khan et al., 1991 and Khan et al., 1992). A model for meso-scale plastic deformation of clayey rocks has also been developed by Shen et al. (2012).

This work aims to provide a computationally efficient approach to simulate complex pressure sensitive elastoplastic deformation of porous rocks coupled together with their mechanical and hydraulic driven damages. Due to the inherent microstructural anisotropy of the porous rocks, an anisotropic CDM framework is developed to take into account the interaction between mean stress and shear stresses in the dynamic fracture of rocks. It is shown that this developed theory performs quite well in capturing experimental data. To show implementation of this method, we considered a hydraulic fracturing problem as an example. Hydraulic fracturing is extensively utilized to enhance oil and gas productions in low permeability formations. A primary difficulty in hydraulic fracturing problems comes from the coupling of the fluid flow inside the fracture and the rock deformations, which provide crack openings. Unfortunately, analytical solutions are limited to simple geometries and limiting assumptions such as homogenous or isotropic medium. In the general case, solutions for fluid-driven fractures are tremendously difficult to construct even for simple geometries (Adachi et al., 2007). This difficulty is due to moving boundary conditions, non-linearity of the governing equation for fluid flow in fractures, high gradient of displacement near the fracture tips, and non-locality of the solution. Non-linearity comes from the fact that fracture permeability is a cubic function of the fracture width. Non-locality means that the fracture opening at one point is a function of fluid pressure at another point along the fracture. To address the above-mentioned challenges, several numerical methods using the finite element analysis (Dahi Taleghani and Olson, 2011 and Sarris and Papanastasiou, 2012) and the boundary element methods (Cleary et al., 1983 and Olson and Dahi Taleghani, 2009) have been proposed in the literature to model the hydraulic fracturing propagation. Methods like cohesive interface models have recently used in the literature to model failure propagation in the porous materials like hydraulic fracturing in rocks (Sarris and Papanastasiou, 2012), however, cohesive models only consider the solid part of the stress. In other words, the cohesive element technique is based on effective stress calculations for failure initiation and propagation. Therefore these models fail to predict changes in rock poroelastic properties like Biot coefficient and Biot modulus. Additionally, the cohesive element method encounters difficulties in situations involved intersecting discontinuities like intersection of advancing hydraulic fracture with pre-existing natural fractures. The method proposed in this paper would overcome some limitations of current methods for modeling hydraulic fracturing problems like remeshing, mesh dependency and numerical instability. Additionally, the proposed model may include the effect of micro natural fractures and their degree of digenesis in simulating hydraulic fracture propagation under real conditions.

The paper is designed as follows: in the next section, the underlying kinematics of the theory is discussed briefly; Section 3 gives a short review of isotropic and anisotropic poroplastic deformation in rocks and the effect of confining stresses on rock's mechanical and hydraulic properties. Section 4 describes different mechanisms of damage in rocks which are later correlated to the constitutive material properties in Section5 using CDM framework. Section 6 outlines the computational aspects for the developed method; and Section 7 presents the numerical simulation results and compares them with the available experimental measurements. It should be noted that the present work provides a phenomenological approach to formulate the inelastic and damage mechanisms in porous rocks. The thermodynamic consistency requirements for plasticity and damage models have been discussed elsewhere, e.g. (Shao et al., 2004 and Voyiadjis et al., 2011).

KINEMATICS

The strain formulation in this work will follow the small strain kinematics. The additive decomposition of the elastic, plastic and damage strain rates is then adopted here to provide the flexibility of defining separate constitutive relations for each stage of the elastic, plastic and damage deformations. Thus, the total strain rate tensor, $\dot{\varepsilon}_{ij}$, is decomposed into the elastic $\dot{\varepsilon}_{ij}^e$, plastic $\dot{\varepsilon}_{ij}^p$, and damaged $\dot{\varepsilon}_{ij}^D$, strain rate components as:

$$\dot{\epsilon}_{ij} = \dot{\epsilon}_{ij}^e + \dot{\epsilon}_{ij}^p + \dot{\epsilon}_{ij}^D, \tag{1}$$

The constitutive relations for these strain components are formulated in the following Sections. The additive decomposition is also used in the development of the return mapping solution algorithms, as discussed by Simo and Ortiz (1985). In the mechanics of porous medium, it is convenient to further decompose the elastic, $\dot{\varepsilon}_{ij}^{e}$, plastic, $\dot{\varepsilon}_{ij}^{P}$, and damaged, $\dot{\varepsilon}_{ij}^{D}$, strain rates into their dilatational ($\dot{\Theta}^{\#} = \dot{\varepsilon}_{kk}^{\#}$) and deviatoric ($\dot{\gamma}_{ij}^{\#}$) components as follows:

$$\dot{\varepsilon}_{ij}^{\#} = \dot{\gamma}_{ij}^{\#} + \frac{1}{3}\dot{\Theta}^{\#}\delta_{ij}, \qquad (2)$$

Using the formulation approach of Eq. (2), the underlying deformation mechanisms of a porous rock can be readily explained when the dilatational and shear deformations are separately formulated.

POROELASTICITY AND POROPLASTICITY OF ROCKS

In this Section, the isotropic and anisotropic poroelasticity frameworks are briefly discussed and then the inelastic deformation of rocks is considered. Furthermore, the pressure sensitivity of deformation mechanisms in porous rocks, which depends on the rock's compressibility, is investigated in this Section.

In the case of isotropic poroelastic rock, the main mechanisms in interaction between the fluid and the porous rock are rock dilation and fluid diffusion. First, an increase in pore pressure induces a dilation of the rock, and second a compression of the rock causes increase of pore pressure, in the undrained condition. However, in the drained condition, excess pore pressure induced by compression of the rock dissipates and secondary deformation of the rock takes place (Detournay and Cheng, 1993). Biot (1941) developed the general theory of three dimensional consolidation by solving coupled diffusion and elasticity equations, and later added temperature effects into his theory (Biot, 1956a and Biot,

1956b). Completing this theory, Rice and Cleary (1976) developed constitutive equations for linear, isotropic, fluid-infiltrated porous media

$$\epsilon_{ij} = \frac{1}{2\mu}\sigma_{ij} - \left(\frac{1}{2\mu} - \frac{1}{3K}\right)\Sigma\delta_{ij} + \frac{1}{3K}\alpha p\delta_{ij},$$

$$\zeta = \frac{1-v}{2\mu(1+v)}\alpha\left(3\Sigma + \frac{p}{B}\right), \qquad (3)$$

where $\Sigma = \frac{1}{3}\sigma_{kk}$ is the mean stress part of the applied stress, δ_{ij} is the Kronecker delta, ε_{ij} and σ_{ij} are the strains and stresses in the solid matrix and p is the fluid pore pressure. Here ζ (the increment of fluid content) is defined as the mass of pore fluid per unit bulk volume. Note that there are four independent material constants in the above equations: shear modulus μ, drained Poisson's ratio v, Biot–Willis coefficient α, and Skempton's coefficient B. The first equation relates strain, ε_{ij}, Cauchy stress, σ_{ij}, and pore pressure, p; and the second equation relates the changes in the fluid mass per unit volume to the first invariant of stress tensor, 3Σ.

Reformulation of the above equations for anisotropic rocks are developed based on Biot (1956b) theoretically by Thompson and Willis (1991); and later, Cheng (1997) proposed a methodology to experimentally measure anisotropic poroelastic properties. The effective stress tensor, i.e. $\dot{\sigma}'_{ij}$, is given by:

$$\sigma'_{ij} = M^h_{ijkl}\epsilon_{kl} - \alpha_{ij}p,$$

$$p = M(\zeta - \alpha_{ij}\epsilon_{ij}), \qquad (4)$$

where α_{ij} is the Biot coefficients tensor and M is the Biot modulus. One may note that unlike the isotropic case, the Biot coefficient is not scalar. Symmetry of the stress tensor requires that the Biot coefficients tensor be symmetric; therefore, it is expected to have six independent parameters. M^h_{ijkl} is the elastic stiffness tensor for the homogenized porous material, where the symmetry of stress and strain tensors reduces its independent parameters to 21 and could be related to the drained elastic tensor through the following relationship:

$$M^h_{ijkl} = M^d_{ijkl} - M\,\alpha_{ij}\alpha_{kl}, \qquad (5)$$

where M^d_{ijkl} is the "drained" stiffness tensor. It is worth noting that in the case of anisotropic poroelasticity, the deviatoric part of the applied stress can also affect the pore pressure. The following relations exist between the material parameters (Shao et al., 2004 and Lu et al., 2013):

$$\alpha_{ij} = \delta_{ij} - \frac{1}{3\|K\|} C^d_{ijkk},$$

$$M = \|K\| \left[\left(1 - \frac{1}{9\|K\|} C^d_{iijj}\right) - \phi\left(1 - \frac{\|K\|}{K_f}\right) \right], \qquad (6)$$

where $C^d_{ijkl} = (M^d_{ijkl})^{-1}$ is the fourth order elastic compliance tensor for drained conditions, $\|K\|$ is the norm of solid bulk modulus matrix, to be defined later, K_f is the fluid bulk modulus, and is the porosity.

In the case of inelastic deformation of rocks, it is assumed in this work that the rocks plastic constitutive behavior is isochoric. In other words, the plastic strain in rocks is only generated by the deviatoric part of the applied stress. The inelastic dilatational deformation of porous rocks is then captured through dilatational damage mechanisms such as void nucleation and growth. Thus, the isochoric plastic deformation assumption results in $\dot{\Theta}^p = 0$ while constitutive relations for the volumetric damage strain, i.e. $\Theta^d = \frac{1}{3}\varepsilon^d_{kk}$, will account for the volumetric inelastic changes and compressibility of the rock. The elastic shear strain rate is defined based on Hook's law as follows:

$$\dot{\gamma}^e_{ij} = \frac{1}{2\mu}\dot{S}_{ij}, \tag{7}$$

where μ is the shear elastic modulus, dot indicates the time derivative (rate), $S_{ij} = \sigma_{ij} - \frac{1}{3}\Sigma\delta_{ij}$ is the deviatoric stress tensor, σ_{ij} is the Cauchy stress tensor, and $\Sigma = \frac{1}{3}\sigma_{kk}$ is the mean stress. The rate-independent-isothermal plastic response of the rock is prescribed through a Johnson–Cook's (Johnson and Cook, 1985) type relation as follows:

$$|\tau|^u = \left[|\tau|^u_P + B|\epsilon_p|^n\right], \tag{8}$$

where $|\tau|^u_P$ is the pressure sensitive von-Mises strength, which is defined later in Eq. (9), $|\epsilon p|$ is the accumulative equivalent plastic strain; and B (MPa) and n are two material parameters which are available for a wide set of rocks and ceramics in the literature (Johnson and Cook, 1985 and Steinberg, 1996).

The mean stress has a significant effect on rock's constitutive behavior; then the state of the applied stress triaxility significantly affects the deformation mechanisms in porous rocks. The rocks exhibit lower strength in the case of applied tensile pressures; while they have superior strength when they are subjected to the compressive pressure fields. On the other hand, their mechanical properties, such as shear and bulk moduli, are affected by the state of the applied pressure. Here we propose empirical relations to account for pressure effect on the (i) strength and (ii) elastic properties of rocks in the following.

Mean Stress Effect on Rock's Strength

The sensitivity of the rock's ultimate strength, i.e. $|\tau|_P^u$, to the confining pressure field can be defined based on the Homquist and Johnson (HJ) model (Holmquist and Johnson, 2005 and Holmquist and Johnson, 2008). Let's define the effective mean stress acting on a rock element by $\Sigma' = \Sigma - p$. An intact rock shows zero strength to tensile effective pressures greater than the rock dilatational tensile strength, i.e. T (MPa), viz. $|\Sigma'| > |T|$. The strength of the rock is assumed to be linearly increased from zero to $|\tau|_0^u$ when the mean stress is changed from tensile $-T$ up to the compressive $|\Sigma'|_0$ pressure. In the case of any compressive pressure greater than $|\Sigma'|_0$, i.e. $|\Sigma'| > |\Sigma'|_0$, the strength is given by:

$$|\tau|_P^u = |\tau|_0^u + (|\tau|_{max}^u - |\tau|_0^u)[1 - \exp(-a(\Sigma' - |\Sigma'|_0))], \quad \text{for} \quad \Sigma' > |\Sigma'|_0, \tag{9}$$

where $|\tau|_{max}^u$ is the maximum achievable rock's strength at high compressive pressures and the constant a is given by:

$$a = \frac{|\tau|_0^u}{(|\tau|_{max}^u - |\tau|_0^u)(|\Sigma'|_0 + T)}. \tag{10}$$

The material parameters in Eq. (10) are readily calibrated with respect to the triaxial test data, and Eqs.(9) and (10) provide a simple and efficient representation of the pressure sensitivity for the rock's strength.

Pressure Effect on Rock's Mechanical Properties

The coupling between the rock's constitutive response and the mean stress changes is formulated by assuming that the undamaged shear, $\bar{\mu}$ and bulk, $\bar{K}$ moduli of the rock obey the following correlations

$$\bar{\mu} = f_p \mu_0 \quad \text{and} \quad \bar{K} = f_p K_0, \qquad (11)$$

where E_0 and K_0 are the reference tensile and bulk elastic moduli at zero mean stress, respectively, and f_p is a function to capture the effect of the pressure variations on the elastic material properties and it is represented by a linear curve fitting to the experimental data:

$$f_p = 1 + \frac{\frac{E_1}{E_0} - 1}{\Sigma'_1} \Sigma', \qquad (12)$$

where E_1 is the tensile modulus of the rock at Σ'_1 which is extrapolated from experimental data.

DAMAGE MECHANICS IN POROUS ROCKS

The damage mechanisms in rocks are categorized based upon loading, fluid content and environmental conditions. These can be divided into three basic classes: (i) Shear damage: the applied stresses in drained conditions will result in micro-crack formation and propagation within the rock's skeleton. A fracture mechanics based CDM model is discussed in Section 4.1 to formulate the shear damage. (ii) Dilatational damage: Once the pore pressure provides a tensile effective pressure field, i.e. Σ', greater than the rock's tensile pressure strength, T, volumetric damages occur and degrade the bulk properties through nucleation and coalescence of voids. The dilatational damage is studied in Section 4.2, in which a fracture mechanics based model is proposed for the volumetric damages. (iii) Corrosion Damage: Even under low stress amplitudes, the corrosive mechanisms deteriorate the mechanical properties of the rocks. This class of damage has not been considered in this work and generally occurs in unconsolidated formations in the form of channelization (Rostami and Dahi Taleghani, 2014).

As discussed in detail by Lemaitre and Chaboche (1990), at least two damage parameters are required to accurately capture the damage mechanism in materials. In this work, only the first two of the abovementioned damage categories are considered to study the fracture in porous rocks. These damage models are later linked to the constitutive responses of drained and undrained rocks in Section 5 via CDM framework.

In the context of CDM, the density of the micro-flaws are represented with a damage parameter as discussed in detail by Kachanov (1958), and latter generalized for the healing effect by Voyiadjis et al. (2012a). Following (Sayers and Kachanov, 1991 and Lubarda and Krajcinovic, 1993), the state of anisotropic damage is introduced by a second rank damage tensor, $d^{\#}_{ij}$. To simplify the formulation it is assumed that the anisotropic damages are mapped onto the three principal directions. Then the damage tensor reads:

$$d_{ij}^{\#} = \begin{bmatrix} D_1^{\#} & 0 & 0 \\ 0 & D_2^{\#} & 0 \\ 0 & 0 & D_3^{\#} \end{bmatrix},$$

(13)

where superscript '#' will be replaced by "s" and "d" to indicate shear and dilatational damage, respectively, and $D_i^{\#}$ is the damage parameter in principal directions, to be defined in the following subsections.

Mechanics of Shear Damage in Rocks

Rocks contain inherent micro-cracks based on their mineralogical compositions and history of loading that induces an inherent anisotropy in the fracture process zone of rocks during a hydraulic fracturing. Due to these anisotropy effects, the rock skeleton experiences 3D state of stress, even under pure mean stress loading conditions. Basically, the deviatoric part of the applied stress is the major driving force for the nucleation and propagation of micro-crack and this class of damage mechanisms deteriorate the rock's shear modulus. Thus, the shear damage parameter, D_i^s, is prescribed using a fracture mechanics based approach in which the deviatoric part of the applied stress controls the micro-cracking. The micro-cracks, within the Representative Volume Element, are mapped into the principal directions and then the average rate of their propagations are considered (Nemat-Nasser and Horii, 1982 and Shojaei et al., 2013). The shear damage parameter is introduced based on works by Sayers and Kachanov, 1991 and Lubarda and Krajcinovic, 1993, who have developed fracture mechanics based CDM models:

$$D_i^s = \left(\frac{\hat{a}_i - a_{i0}}{a_i^c - a_{i0}} \right)^q \quad \text{with } i = 1, 2, \text{ and } 3,$$

(14)

where $\hat{a}_i = \frac{1}{m}\sum_{k=1}^{N} \hat{a}_{iK}$ is the averaged micro-crack length in the ith direction and is called "the representative micro-crack", q is a material parameter that calibrates the rate of the fracture, a_{i0} is the initial length of the micro-flaws in the i th direction, and a^c_i is the critical crack length at which instable fracture occurs in the i th direction. The shear damage tensor is then correlated to the evolution of micro-cracks in each of the three principal directions.

The constitutive relationship for a_{i0} is introduced based upon the state of the plastic deformation in rocks where the deviatoric part of the applied stress induces inelastic strains. This approach is adopted because the plastic strain increment, i.e. $\Delta\varepsilon_{ij}^p$, is available as a field output in most of the commercial FEA packages and the developed model can be readily implemented through FEA software. The micro-crack evolution law then reads:

$$\hat{a}_i = a_{i0} + \chi_5(a^c_i - a_{i0})\sum_{k=0}^{N}\frac{|\Delta\epsilon_p|_k}{\epsilon_f}, \tag{15}$$

where k indicates the load step, and χ_5 is the proportionality factor, to be defined later in Eq. (19). The equivalent plastic strain increment, $|\Delta\epsilon_p|$, and the final fracture strain ϵ_f are defined as follows:

$$|\Delta\epsilon_p| = \sqrt{\frac{2}{3}\Delta\epsilon_{ij}^p \Delta\epsilon_{ij}^p},$$
$$\epsilon_f = \{d_1 + d_2\exp(-d_3\lambda)\}, \tag{16}$$

where d_1, d_2, and d_3 are material constants (available from the literature, e.g. (Addessio and Johnson, 1990 and Steinberg, 1996)) and λ denotes the state of the stress triaxiality, i.e. $\lambda=\Sigma'/|\tau|$.

The microcrack initiation criterion is introduced based upon the fracture mechanics of rocks herein. Nemat-Nasser and Hori have introduced the concept of sliding cracks in which the normal and shear stresses, applied to the crack surface, are used to predict the crack advances in mode-I and mode-II fractures (Nemat-Nasser and Horii, 1982). This concept is generalized by Shojaei et al. who proposed a CDM model for dynamic fracturing of polycrystalline materials (Shojaei et al., 2013), where the crack propagation is linked to the state of the applied stress triaxiality.

Excessive fluid pressure inside the cracks tried to open them or at least reduce the contact between crack faces, and further propagated in Mode-I, -II, -III or combination as mixed-mode conditions. When the applied mean stress is compressive, the frictional sliding between the crack surfaces plays an important role (Nemat-Nasser and Horii, 1982). Considering the shearing nature of rock failure under large confining stresses, the crack initiation criterion in this case can be given by Shojaei et al. (2013):

$$\sqrt{2\pi \hat{a}_i} \chi_3 \left[-(\chi_4 - 1)\Sigma' + \chi_5 \frac{(\chi_4 + 1)}{\sqrt{3}} |\tau| \right] - K_i^{IC} = 0, \tag{17}$$

where Σ' is the effective mean stress, $|\tau| = \sqrt{\frac{3}{2} S'_{ij} S'_{ij}}$ is the effective shear stress, $S'_{ij} = \sigma'_{ij} - \Sigma' \delta'_{ij}$ is the effective deviatoric part of applied Cauchy stress, K_i^{IC} is a representative of the rock fracture toughness at the ith direction, χ_5 is a proportionality factor to take into account the interaction between propagated micro-cracks. Parameters χ_3, and χ_4 are material parameters to account for the frictional sliding effect and defined as:

$$\chi_3 = \frac{(1+\mu_c^2)^{1/2} - \mu_c}{\sqrt{3}},$$

$$\chi_4 = \frac{(1+\mu_c^2)^{1/2} + \mu_c}{(1+\mu_c^2)^{1/2} - \mu_c},$$

(18)

and proportionality factor is defined by:

$$\chi_5 = \begin{cases} \chi_6\left(\frac{\hat{a}_i}{a_i^c}\right), & \text{for } \hat{a}_i < a_i^c \\ \chi_6, & \text{for } \hat{a}_i \geq a_i^c, \end{cases}$$

(19)

where μ_c is the Coulomb's friction coefficient for in-contact crack surfaces. The parameter χ_6 is a material constant to calibrate the effect of micro-cracks interactions in the unstable fracture process. The material parameters in the damage criterion are controlled by the microstructure and fabric of the rock and derived from experimental measurements as follows. It is assumed that a_{i0} is equal to the average grain size in each of the principal directions, and K^{IC}_i equates to the Mode-I critical stress intensity factor. The remaining two material constants, i.e. χ_6 and a^C_i, are directly determined from the experimental data. These two sets of material parameters are calibrated based upon the triaxial test data, in which the damage initiation and final failure points are correlated to these parameters as follows (Shao et al., 2004):

$$(\sigma_3 - \sigma_1)_{\text{peak}} = \frac{3K_i^{IC}}{\chi_6 \sqrt{\pi a_i^c}} - \frac{3}{\chi_6}\sigma_3,$$

$$(\sigma_3 - \sigma_1)_{\text{init}} = \frac{3K_i^{IC}}{\chi_6\left(\frac{a_i^c}{a_{i0}}\right)\sqrt{\pi a_{i0}}} - \frac{3}{\chi_6\left(\frac{a_i^c}{a_{i0}}\right)}\sigma_3,$$

(20)

where subscripts "peak" and "init" denote respectively the peak load and deviation from linear response in the triaxial test.

Due to the strong stress singularity at the crack tip, a stress based crack propagation criterion is not applicable in the case of post fracture stress analysis of rocks. Thus, a fracture mechanics based criterion is developed herein to remove the stress singularity effect from the problem. The mixed-mode fracture criterion is given by

$$f_G^D(G_i) = \left(\frac{G_I}{G_{Ic}}\right)^2 + \left(\frac{G_{II}}{G_{IIc}}\right)^2 + \left(\frac{G_{III}}{G_{IIIc}}\right)^2 + \left(\frac{G_I G_{II}}{G_{Ic} G_{IIc}}\right) + \left(\frac{G_{II} G_{III}}{G_{IIc} G_{IIIc}}\right) + \left(\frac{G_I G_{III}}{G_{Ic} G_{IIIc}}\right).$$

(21)

where G_{ic} with i = I, II and III are critical energy release rates, and fracture dissipative energy release rates, i.e. G_i, are computed based upon the traction–displacement responses of the interfacial medium between two crack faces:

$$G_i = \int_0^{\delta_i} \tau_i d\delta_i \text{ with } i = I, II, \text{ and } III,$$

(22)

where δ_i is the crack opening displacement and τ_i is the traction. One may note that Eq. (21) takes into account the coupling effect between the three fracture modes.

Dilatational Damage Mechanics of Rocks

The applied tensile mean stresses in a hydraulic fracturing, due to the fluid pore pressure effect, may result in volumetric damages nucleation and coalescence that eventually affect the bulk properties of the rocks. The dilatational inelastic volume is prescribed here by:

$$V^d = V_0 + (V_c - V_0) \sum_{i=1}^{N} \frac{\Delta\Theta^d}{\epsilon^f},$$

(23)

where V^d is a representative value for the inelastic volume changes, V_0 indicates the initial volume of porosities, V_c is the critical porosity volume fraction in which the unstable failure occurs, and $\Delta\Theta^d$ is the dilatational inelastic strain increment. It should be noted that there is a clear distinction between the dilatational inelastic,

i.e. $\Delta\Theta^d = \frac{1}{3}\Delta\varepsilon^d_{kk}$, and plastic, $\dot{\Theta}^p = \Delta\varepsilon^p_{kk} = 0$, strains in this work. As already discussed, the dilatational plastic deformation is zero due to the isochoric assumption for the rock skeleton's plastic behavior. On the other hand, due to the presence of anisotropic damage mechanisms in porous rocks, there is a dilatational inelastic strain due to the failure of skeleton structures. Thus, excessive plastic deformation may result in failure of the rock's solid skeleton and produce higher porosity. The inelastic dilation $\Delta\Theta^d$ is then computed based on anisotropic damage mechanisms in porous rocks. This point is particularly important to avoid any confusion between isochoric assumption in Eq. (7) for incompressible plastic deformation, which is valid only for the rock skeleton, and utilizing an inelastic dilation $\Delta\Theta^d$ in Eq. (23) that is related to damage mechanisms in porous rocks.

While the dilatational damage in this work is controlled by the anisotropy of the induced shear damages, one may further constraint the dilatational damage evolution by utilizing an initiation criterion for the void nucleation. The void nucleation criterion can be defined based upon a stress based criterion, as follows:

$$\chi_7 \Sigma' - \Sigma_N = 0, \tag{24}$$

where Σ_N is the required mean stress to initiate the volumetric damages, available in the literature (Steinberg, 1996), and χ_7 takes into account the coalescence effects as defined by:

$$\chi_7 = \begin{cases} \chi_8(\frac{V}{V_c}), & \text{for } V < V_c \\ \chi_8, & \text{for } V \geqslant V_c \end{cases}, \tag{25}$$

where χ_8 is a material parameter that controls the interaction between propagating voids. The dilatational damage parameter is then introduced by:

$$D_i^d = D^d I_i = \left(\frac{V^d - V_0}{V_c - V_0}\right)^q I_i, \tag{26}$$

where I_i is the unity vector.

CONTINUUM DAMAGE MECHANICS (CDM) FORMULATION FOR HYDRAULIC FRACTURING

In Section 4 the damage mechanisms in rocks are formulated. In this section the developed shear and dilatational damage parameters are correlated to the constitutive behavior of rocks. Generally, in the context of CDM, there are two approaches which are utilized to establish the relations between damaged and undamaged material properties. These two approaches are equivalence of strain energy densities and equivalence of strain between the damaged and fictitious effective configurations (de Sciarra, 1997, Voyiadjis and Kattan, 2006, Desmorat et al., 2007, Voyiadjis et al., 2011, Voyiadjis et al., 2012a and Shojaei et al., 2013). Here, the CDM approach is adopted to formulate the anisotropic damage process in rocks. First the shear modulus is linked to the shear damage parameter as follows:

$$\mu_i = \bar{\mu}(I_i - D_i^s)^2, \tag{27}$$

where $\bar{\mu}$ is the elastic shear modulus for an undamaged drained rock, as defined by Eq. (11). Furthermore, due to the isotropy of the dilatational damage, the following correlation holds:

$$K_i = \bar{K}(I_i - D_i^d)^2, \tag{28}$$

where $\bar{K}$ is the bulk modulus for the undamaged drained rock, as defined by Eq. (11). All other material parameters are computable by correlations between elastic constant. The components of the drained elastic moduli tensor, M^d_{ijkl}, is then computed based upon updated damaged values of μ_i and K_i in each time step.

Furthermore, to include the effect of the dilatational damage on the pressure sensitivity of the strength of the system, the following damage evolution laws are applicable:

$$T = \bar{T}(1 - ||D^d||)^2;$$
$$|\Sigma'|_0 = |\bar{\Sigma}'|_0 (1 - ||D^d||)^2;$$
$$|\tau|^u_{max} = |\bar{\tau}|^u_{max}(1 - ||D^s||)^2;$$
$$|\tau|^u_0 = |\bar{\tau}|^u_0 (1 - ||D^s||)^2; \qquad (29)$$

where " $||\ ||$ " indicates the norm of the tensor, and the over-bar parameters indicate the respective parameter for the intact rock (undamaged). In Eq. (29) the dilatational parameters, i.e. T and $|\Sigma'|_0$, and strength value $|\tau|^u_0$, and maximum strength $|\tau|^u_{max}$, are degradated through shear/dilatational damage relation.

CDM Model for Undrained Condition

The damage mechanisms at the presence of a freely moving fluid in a porous rock will affect the mechanical and hydraulic responses of the medium. The skeleton of porous rocks is constituted from micro- to macro-scale unit cells and upon failure of each of these individual cells the pore pressure, fluid content, as well as the load carrying capacity of the skeleton are altered. To account for these changes the material parameters of a porous rock are correlated to the damage parameter within the CDM framework as follows:

$$M = \bar{M}(1 - ||D^s||)(1 - ||D^d||),$$
$$\alpha_{ij} = \bar{\alpha}_{ik}(\delta_{kl} - D^s_{kl})(\delta_{lj} - D^d_{lj}), \qquad (30)$$

where the over-bar denotes the undamaged parameters for an intact rock. It is worth noting that the proposed CDM models in Eq. (30) take into account the coupling effect between the dilatational and shear damage mechanisms in deteriorating the poroelastic material properties of the rock medium. In this work, the damaged stiffness tensor for a drained rock is utilized to capture the poroelastic damage, via Eq. (6).

COMPUTATIONAL ASPECTS

The proposed failure model is introduced into a commercial FEA package (ABAQUS) through user-defined subroutines, i.e. USDFLD and UMAT. The dynamic implicit integration scheme together with enhanced hourglass meshes is utilized to avoid mesh dependent results. Two developed damage variables, i.e. d^s_{ij} and d^d_{ij}, are updated incrementally at each material point as field variables and the updated values are utilized to reduce the elastic shear and bulk moduli, i.e. μ_i and K_i, respectively. Once the reduced damaged elastic moduli, i.e. μ_i or K_i, in one of the integration points reaches a certain threshold, viz. 0.6 of the intact moduli, the corresponding element is removed from the model by setting a small value for its elastic modulus. The FEA implementation flowchart is depicted in

Fig. 1, where F^p F^d_s and $F^d_{\Sigma'}$ respectively indicate plastic, shear damage and volumetric damage criteria. The return mapping algorithm is chosen here to compute the increments of the plastic and damage parameters in the FEA model. The details for the return mapping and implicit integration schemes can be found in (Simo and Hughes, 1997 and Voyiadjis et al., 2012b).

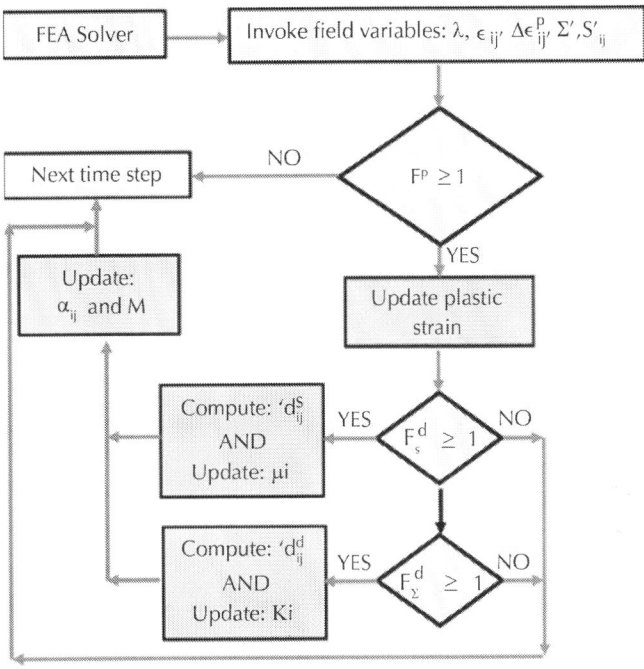

Figure 1: Outlines of the numerical method for computing plastic strain, shear and volumetric damages.

One may notice that upon failure of an element, the progressive fracture is controlled via the quadratic fracture mechanics criterion. This approach ensures the stress singularities, associated with sharp edges, do not affect the simulation results.

RESULTS AND DISCUSSION

The performance of the developed framework is examined in this section. The test data, used in what follows, are performed on a brown colored Sandstone and reported by Khazraei (1995). Fig. 2 depicts the pressure sensitivity effect on the rock's strength. As confirmed by the experimental results, rocks show superior strength under compressive pressure fields; while they show little resistance when they are subjected to the tensile pressure field, like the hydraulic fracturing case. Table 1 summarizes the material parameters that are used in computations of pressure sensitive strength. Note that in Fig. 2 the positive Σ' denotes compressive effective pressure field.

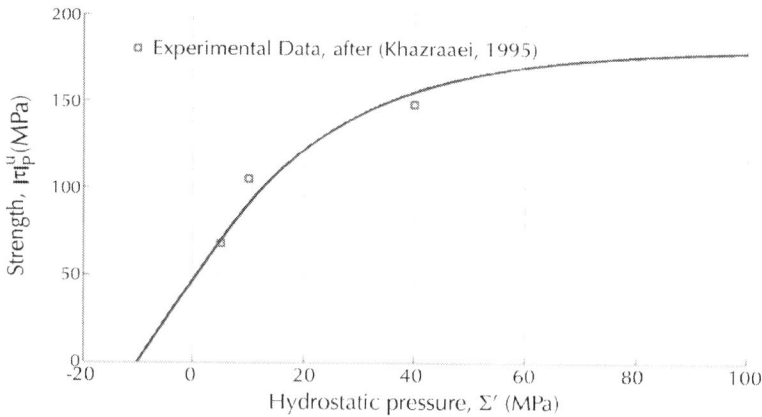

Figure 2: Mean stress effect on the strength of the rock, experiments are after (Khazraei, 1995).

Table 1: Parameters for the pressure sensitive strength, Eq. (10)

$|\tau|^u_{max}$ (MPa)	$|\tau|^u_0$ (MPa)	$|\Sigma'|_0$ (MPa)	T (MPa)
180	70	5	−10

The pressure sensitivity of material properties of rocks are studied in Fig. 3, where the computed elastic tensile modulus (from shear and bulk moduli in Eq. (12)) is portrayed in comparison with experimental data points. It is noted that the variation of elastic tensile modulus is not significant; although, the proposed empirical relation in Eq. (12) performs quite accurately to capture these changes. Table 2 shows the material parameters that are used in Fig. 3 simulation.

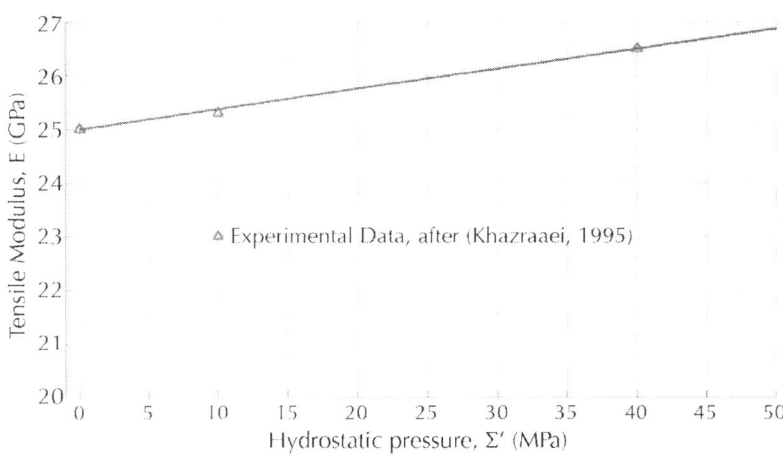

Figure 3: Mean stress effect on tensile modulus of the rock, experiments are after (Khazraei, 1995).

Table 2: Parameters for the pressure sensitive tensile modulus, Eq. (12)

E_1 (GPa)	Σ'_1 (MPa)	E_0 (GPa)	v
26.5	40	25	0.3

The elasto-plastic deformation of Sandstone has been experimentally investigated by Khazraei (1995) where triaxial tests were carried out to study the deformation mechanisms of rocks under various pressure effect. As depicted in Fig. 4, the proposed elastoplastic formulation correlates with the experimental data. Table 3 indicates the material parameters for elastoplastic simulation and shear and dilatational damage computations. Fig. 5 illustrates the performance of the shear and dilatational damage model to simulate the tensile elastic modulus degradation. In Fig. 5, the shear and bulk moduli are updated with respect to the state of the plastic strain and then they are utilized to compute the tensile modulus. The damage simulation in Fig. 5, shown with solid line, follows the experimental data points quite accurately. It is worth noting that in Table 3, some parameters such as G_{Ic}, G_{IIc}, G_{IIIc}, Σ_N, V_0, a_{10}, μ_c are obtainable from experimental measurements, while other parameters such as χ_6, χ_8, B, n, q, d_1, d_2, and d_3 are determined from curve fitting techniques.

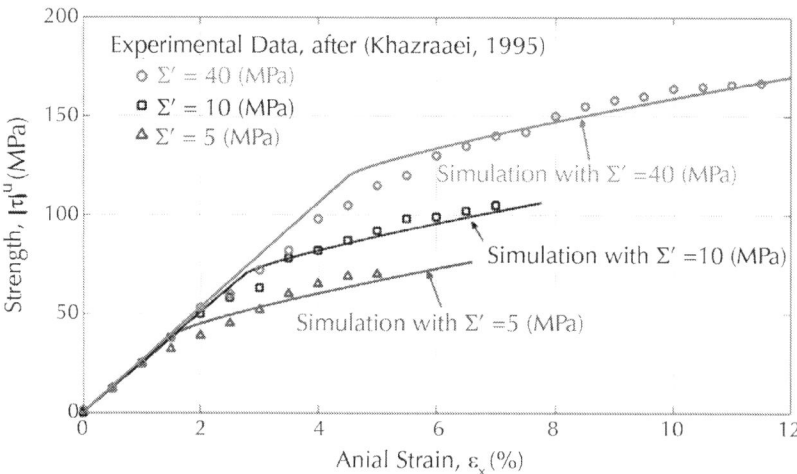

Figure 4: Pressure sensitive stress–strain responses of porous rocks, experiments are after (Khazraei, 1995).

Table 3: Parameters for the pressure sensitive stress–strain and shear damage responses, Eqs. (8), (15) and (18)

B(MPa)	n	χ_6	χ_8	a_{10} (mm)	a_1^c (mm)	V_0 (mm³)	V_c (mm³)	μ_c	Σ_N (MPa)	q	d_1(%)	d_2(%)	d_3	G_{Ic} (kPa m)	$G_{IIc} = G_{IIIc}$ (kPa m)
10	0.8	1	1	0.1	1	1e–3	1	0.65	45	2	3.5	0.5	0.1	0.224	0.15

A Continuum Damage Failure Model for Hydraulic Fracturing ...

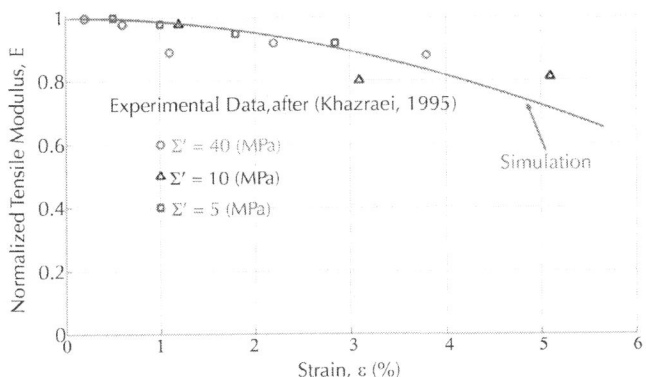

Figure 5: Variation of elastic tensile modulus of a rock due to accumulative shear and dilatational damages, experiments are after (Khazraei, 1995).

It is worth noting that the simulation in Fig. 5 is obtained by updating the tensile modulus based upon damaged shear and bulk moduli. In Fig. 6 the crack initiation criterion is used to obtain the state of effective shear and mean stress at the onset of fracture. The effect of Coulomb's friction coefficient and the proportionality material parameter are also parametrically studied in Fig. 6. It is noted that at the presence of compressive effective pressures, i.e. Σ' > 0, higher shear stresses are required to initiate the fracture; while tensile pressures, i.e. $\Sigma' < 0$, is associated with less shear stress values.

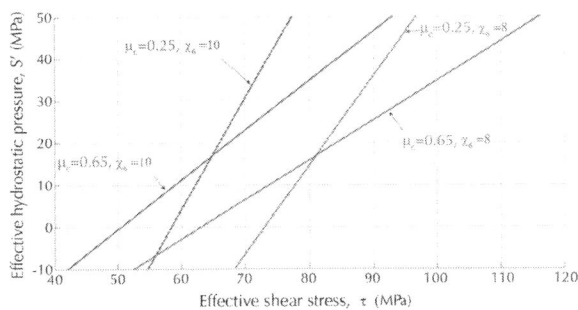

Figure 6: State of effective shear and mean stress (hydrostatic pressure) for the crack initiation, from Eq. (17).

An FEA problem of fluid driven fractures or hydraulic fracture propagation has been considered to show application of the proposed method for practical geomechanical problems. Three domains are considered for the problem, viz. low permeability cap rocks on the top and bottom of the model, hydrocarbon bearing zone, and interfacial domains to represent potential fracture paths, as depicted in Fig. 7. The oil-bearing circular rock slice has a 30 m depth and 400 m radius with a wellbore radius of 0.1 m. The symmetry of the problem only requires modeling a quarter of the problem. Initial conditions of the problem are prescribed by defining an initial geostatic stress field and pore fluid pressure. A depth varying initial void ratio is specified using user subroutine VOIDRI. Gravity loading is specified; and an orthotropic overburden stress state is imposed. Loading and boundary conditions are also applied as follows. The first step is initially achieving the equilibrium state after applying the initial pore pressure and in situ stresses. The second step simulates the hydraulic fracturing stage, where a volume of fluid is being injected into the formation. The fluid flow is injected along the perforation zone in the target formation in the model via the prescribed interfacial medium, see Fig. 7. The interfacial domain, with thickness of 0.1 m shows the fracture process zone in which the damage parameters are updated and elastic moduli degradation occurs. The elastoplastic rock behavior is assigned to the target rock; while confining shale rocks are assumed to behave as linear elastic materials. The duration of the injection stage is 140 s. Following the hydraulic fracture, another transient consolidation analysis is conducted. The injection into the well is terminated, and fluid leakoff from the fracture is allowed to bleed off the fracture fluid pressure. To mimic the real conditions, the fracture surface are assumed to possess a minimum opening as the boundary condition to simulate the behavior of the placed proppant material into the fracture (ABAQUS, 2011). Treatment parameters used for the simulation are listed in Table 4.

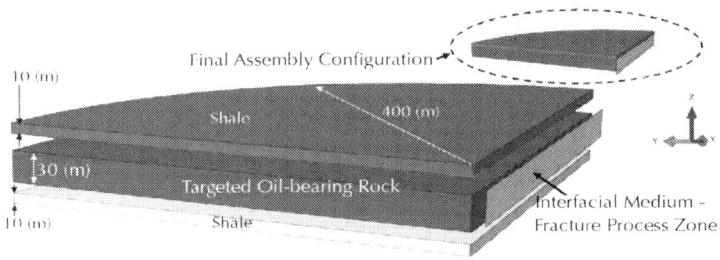

Figure 7: Exploded and assembly views of the FEA problem.

Table 4: Hydraulic fracturing parameters

Viscosity, μ (kPa s)	Injection rate (m^3 s^{-1} m^{-1})	Domain permeability, k (m s^{-1})	In situ effective maximum stress, σ_1 (kPa)	In situ effective intermediate stress, σ_2 (kPa)	In situ effective minimum stress, σ_3 (kPa)	Initial void ratio	Initial pore-pressure (kPa)
0.001	−5e−3	2.418e−11	−15.00e + 3	−12.547e + 3	−8.547e + 3	0.33	354

The pore pressure variation within the interfacial layer in the fracture process zone is shown in Fig. 8 for three step times. It is obvious that the excessive pore pressure induced by fluid injection would result in the damage nucleation or fracture formation. The pressure head is moving along the fracture process zone; and due to the anisotropic failure process the pressure map is uneven across the thickness. It is worthwhile noting that two types of elements have been used in Fig. 8 that are C3D8R for Oil Bearing Rock (see Fig. 7), and C3D8RP for the Interfacial Medium. The later element allows the injection of pore fluid and outputs the pore pressure. The mesh refinement study has also been carried out to check consistency of results with the observed fluid pressure and fracture opening profiles as discussed next.

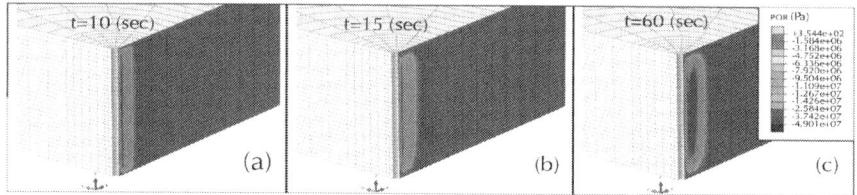

Figure 8: Pore pressure variation within the interfacial rock medium (fracture process zone) is depicted by color variations at (a) 10, (c) 15, and (d) 60 s after injection. The color legend shows the respective pore pressure values in (Pa).

The fracture profile and pressure drop at the fracture front are depicted in Fig. 9. The opening profile agrees with previous studies, such as Sarris and Papanastasiou (2011). In practice this profile can be compared with field or test data to verify the performance of the developed CDM model.

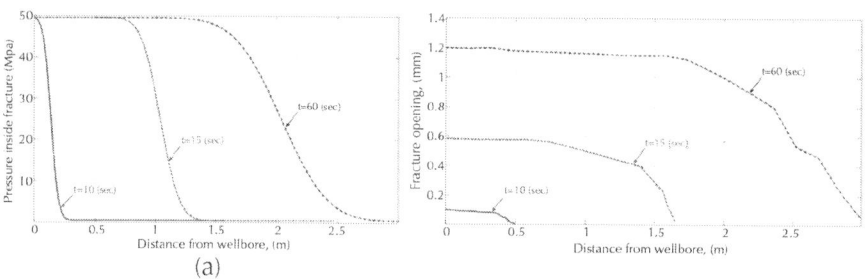

Figure 9: Simulation results for the (a) fluid pressure, and (b) fracture opening profiles.

CONCLUSIONS

The pressure sensitive elastoplastic deformation and damage mechanisms in porous rocks have been studied in detail for both of the drained and undrained conditions. The effect of injected pore fluid in development of fracture surfaces in a hydraulically fractured rock is also investigated in which a CDM based failure model is developed to capture dynamic fracture and post fracture responses of rocks. In the developed approach the fracture is correlated to the microcrack and micro-void nucleation and coalescence. The fracture mechanics of rocks is utilized to develop a physically consistent damage model for porous rocks. In the case of undrained rocks, the poroelastic damage is then captured by updating the Biot Modulus and Biot coefficient based on the damaged stiffness matrix of the drained rock. It is shown that the developed elastoplastic and CDM models correlate well with the available experimental data.

Due to the fact that the conventional FEA elements are utilized in the developed CDM approach, and only the material properties are systematically updated, the developed CDM model computationally outperforms the discrete element analysis approaches, such as cohesive element, for simulating complex hydraulic fracture problems. Furthermore, the fracture mechanics based criterion removes the crack tip's stress singularity issue from the FEA model. This model would provide a robust predictive tool for complex hydraulic fracture simulations, such as interaction of hydraulic and natural fractures. This is a topic of a forthcoming paper by the authors, where the developed CDM model is used to capture the interactions between the natural and hydraulically driven fractures.

ACKNOWLEDGMENTS

This investigation was partially supported by Cooperative Agreement NNX11AM17A between NASA and the Louisiana Board of Regents under contract NASA/LEQSF (2011-14)-Phase3-05. This study was also partially supported by the NSF under Grant number CMMI0900064 and Army Research Office under Grant number W911NF-13-1-0145. ADT also acknowledges the financial supports from Shell Co. during conducting this research.

REFERENCES

1. ABAQUS (2011). Dassault Systemes Simulia Corp. Version 6.11.
2. Adachi, J., Siebrits, E., Peirce, A., Desroches, J., 2007. Computer simulation of hydraulic fractures. Int. J. Rock Mech. Min. Sci. 44, 739–757.
3. Addessio, F.L., Johnson, J.N., 1990. A constitutive model for the dynamic response of brittle materials. J. Appl. Phys. 67 (7), 3275–3286.
4. Aghdam, A.B., Beheshti, A., Khonsari, M.M., 2012. On the fretting crack nucleation with provision for size effect. Tribol. Int. 47, 32–43.
5. Bazˇant, Z.P., Planas, J., 1998. Fracture and Size Effect: In Concrete and Other Quasibrittle Materials. CRC Pres.
6. Bedayat, H., Dahi Taleghani, A., 2014. Interacting double poroelastic inclusions. Mech. Mater. 69 (1), 204–212.
7. Beheshti, A., Khonsari, M.M., 2011. On the prediction of fatigue crack initiation in rolling/sliding contacts with provision for loading sequence effect. Tribol.Int. 44 (12), 1620–1628.
8. Biot, M.A., 1941. General theory of three-dimensional consolidation. J. Appl. Phys. 12, 155.
9. Biot, M.A., 1956a. General solutions of the equations of elasticity and consolidation for a porous material. J. Appl. Mech. 78, 91–96.
10. Biot, M.A., 1956b. Theory of deformation of a porous viscoelastic anisotropic solid. J. Appl. Phys. 27, 459.
11. Brünig, M., Gerke, S., Hagenbrock, V., 2013. Micromechanical studies on the effect of the stress triaxiality and the Lode parameter on ductile damage. Int. J. Plast. 50, 49–65.
12. Budiansky, B., O'Connell, R.J., 1976. Elastic moduli of a cracked solid. Int. J. Solids Struct. 12, 81–97.

13. Cheng, A.D., 1997. Material coefficients of anisotropic poroelasticity. Int. J. Rock Mech. Min. Sci. 34 (2), 199–205.
14. Cleary, M.P., Kavvadas, M., Lam, K.Y., 1983. Development of a fully three-dimensional simulator for analysis and design of hydraulic fracturing. In: SPE/DOE Low Permeability Gas Reservoirs Symposium Denver, Colorado.
15. Cristescu, N., 1988. Viscoplastic creep of rocks around a lined tunnel. Int. J. Plast. 4 (4), 393–412.
16. Dahi Taleghani, A., Olson, J., 2011. Analysis of multi-stranded hydraulic fracture Propagation: an improved model for the interaction between induced and natural fractures. SPE 124884, SPE J. 16(3).
17. de Borst, R., Sluys, L.J., 1991. Localisation in a Cosserat continuum under static and dynamic loading conditions. Comput. Methods Appl. Mech. Eng. 90 (1–3),805–827.
18. de Sciarra, F.M., 1997. A new variational theory and a computational algorithm for coupled elastoplastic damage models. Int. J. Solids Struct. 34 (14), 1761–1796.
19. Desmorat, R., Gatuingt, F., Ragueneau, F., 2007. Nonlocal anisotropic damage model and related computational aspects for quasi-brittle materials. Eng. Fract.Mech. 74 (10), 1539–1560.
20. Detournay, E., Cheng, A.H.-D., 1993. Fundamentals of poroelasticity. In: Hudson, J.A. (Ed.), Comprehensive Rock Engineering: Principles, Practices and Projects, vol. 2. Pergamon Press, Oxford, UK.
21. Doghri, I., Adam, L., Bilger, N., 2010. Mean-field homogenization of elasto-viscoplastic composites based on a general incrementally affine linearization method. Int. J. Plast. 26 (2), 219–238.
22. Hansen, B.L., Carpenter, J.S., Sintay, S.D., Bronkhorst, C.A., McCabe, R.J., Mayeur, J.R., Mourad, H.M., Beyerlein, I.J., Mara, N.A., Chen, S.R., Gray Iii, G.T., 2013. Modeling the texture evolution of Cu/Nb layered composites during rolling. Int. J. Plast. 49, 71–84.

23. Holmquist, T.J., Johnson, G.R., 2005. Characterization and evaluation of silicon carbide for high-velocity impact. J. Appl. Phys. 97 (9), 093502–093512.
24. Holmquist, T.J., Johnson, G.R., 2008. The failed strength of ceramics subjected to high-velocity impact. J. Appl. Phys. 104 (1), 013511–013533.
25. Johnson, G.R., Cook, W.H., 1985. Fracture characteristics of three metals subjected to various strains, strain rates, temperatures and pressures. Eng. Fract. Mech. 21 (1), 31–48.
26. Kachanov, L.M., 1958. Rupture time under creep conditions. Izv. Akad. Nauk SSSR 8, 26–31. Reprinted from: Int. J. Fracture 97, 11–18.
27. Kahirdeh, A., Khonsari, M.M., in press. Criticality of degradation in composite materials subjected to cyclic loading. Compos. Part B: Eng. http://dx.doi.org/ 10.1016/j.compositesb.2013.06.048.
28. Kamrin, K., 2010. Nonlinear elasto-plastic model for dense granular flow. Int. J. Plast. 26 (2), 167–188.
29. Khan, A.S., Xiang, Y., Huang, S., 1991. Behavior of Berea sandstone under confining pressure Part I: yield and failure surfaces, and nonlinear elastic response. Int. J. Plast. 7 (6), 607–624.
30. Khan, A.S., Xiang, Y., Huang, S., 1992. Behavior of Berea sandstone under confining pressure Part II: elastic–plastic response. Int. J. Plast. 8 (3), 209–220.
31. Khazraei, R., 1995. Experimental investigations and numerical modelling of the anisotropic damage of a Vosges sandstone (Doctoral thesis). University of Lille.
32. Krajcinovic, D., Mastilovic, S., 1999. Statistical models of brittle deformation Part I: introduction. Int. J. Plast. 15 (4), 401–426.
33. Kruch, S., Chaboche, J.-L., 2011. Multi-scale analysis in elasto-viscoplasticity coupled with damage. Int. J. Plast. 27 (12), 2026–2039.

34. Lemaitre, J., Chaboche, J.L., 1990. Mechanics of Solid Materials. Cambridge University Press, Cambridge.
35. Li, G., Shojaei, A., 2012. A viscoplastic theory of shape memory polymer fibres with application to self-healing materials. Proc. R. Soc. A: Math. Phys. Eng. Sci. 468 (2144), 2319–2346.
36. Li, G., Xu, W., 2011. Thermomechanical behavior of thermoset shape memory polymer programmed by cold-compression: testing and constitutive modeling. J. Mech. Phys. Solids 59 (6), 1231–1250.
37. Lu, Y.L., Elsworth, D., Wang, L.G., 2013. Microcrack-based coupled damage and flow modeling of fracturing evolution in permeable brittle rocks. Comput. Geotech. 49, 226–244.
38. Lubarda, V.A., Krajcinovic, D., 1993. Damage tensors and the crack density distribution. Int. J. Solids Struct. 30 (20), 2859–2877.
39. Malcher, L., Andrade Pires, F.M., César de Sá, J.M.A., 2012. An assessment of isotropic constitutive models for ductile fracture under high and low stress triaxiality. Int. J. Plast. 30–31, 81–115.
40. Marotti de Sciarra, F., 2012. Hardening plasticity with nonlocal strain damage. Int. J. Plast. 34, 114–138.
41. Moschovidis, Z.A., Mura, T., 1975. Two-ellipsoidal inhomogeneities by the equivalent inclusion method. J. Appl. Mech. 42, 847.
42. Mura, T., 1987. Micromechanics of Defects in Solids. Martinus Nijhoff Publishers.
43. Naderi, M., Kahirdeh, A., Khonsari, M.M., 2012. Dissipated thermal energy and damage evolution of glass/epoxy using infrared thermography and acoustic emission. Compos. B Eng. 43 (3), 1613–1620.
44. Needleman, A., 1987. A continuum model for void nucleation by inclusion debonding. J. Appl. Mech. 54 (3), 525–531.

45. Nemat-Nasser, S., Hori, M., 1993. Micromechanics: Overall Properties of Heterogeneous Materials. Elsevier, Amsterdam.
46. Nemat-Nasser, S., Horii, H., 1982. Compression-induced nonplanar crack extension with application to splitting, exfoliation, and rockburst. J. Geophys. Res. 87, 6805–6821.
47. Nicot, F., Sibille, L., Darve, F., 2012. Failure in rate-independent granular materials as a bifurcation toward a dynamic regime. Int. J. Plast. 29, 136–154.
48. Olson, J., Dahi Taleghani, A., 2009. Modeling simultaneous growth of multiple hydraulic fractures and their interaction with natural fractures. In: HydraulicFracturing Technology Conference, 2009 SPE 119739.
49. Radi, E., Bigoni, D., Loret, B., 2002. Steady crack growth in elastic–plastic fluid-saturated porous media. Int. J. Plast. 18 (3), 345–358.
50. Rice, J.R., Cleary, M.P., 1976. Some basic stress diffusion solutions for fluid saturated elastic porous media with compressible constituents. Rev. Geophys. 14, 227.
51. Rostami, S., Dahi Taleghani, A., 2014. Modeling particle mobilization in unconsolidated formations due to fluid injection. In: 48th US Rock Mechanics/ Geomechanics Symposium. Minnesota, USA.
52. Sarris, E., Papanastasiou, P., 2011. The influence of the cohesive process zone in hydraulic fracturing modelling. Int. J. Fract. 167, 33–45.
53. Sarris, E., Papanastasiou, P., 2012. Modeling of hydraulic fracturing in a poroelastic cohesive formation. Int. J. Geomech. 12 (2), 160–167.
54. Sayers, C.M., Kachanov, M., 1991. A simple technique for finding effective elastic constants of cracked solids for arbitrary crack orientation statistics. Int. J. Solids Struct. 27 (6), 671–680.
55. Shao, J.F., 1998. Poroelastic behaviour of brittle rock materials with anisotropic damage. Mech. Mater. 30 (1), 41–53.

56. Shao, J.F., Henry, J.P., 1991. Development of an elastoplastic model for porous rock. Int. J. Plast. 7 (1–2), 1–13.
57. Shao, J.F., Lu, Y.F., Lydzba, D., 2004. Damage modeling of saturated rocks in drained and undrained conditions. J. Eng. Mech. 130, 733–740.
58. Shen, W.Q., Shao, J.F., Kondo, D., Gatmiri, B., 2012. A micro-macro model for clayey rocks with a plastic compressible porous matrix. Int. J. Plast. 36, 64–85.
59. Shen, W.Q., Kondo, D., Dormieux, L., Shao, J.F., 2013. A closed-form three scale model for ductile rocks with a plastically compressible porous matrix. Mech. Mater. 59, 73–86.
60. Shojaei, A., Li, G., 2013. Viscoplasticity analysis of semicrystalline polymers: a multiscale approach within micromechanics framework. Int. J. Plast. 42, 31–49.
61. Shojaei, A., Li, G., Voyiadjis, G.Z., 2012. Cyclic viscoplastic–viscodamage analysis of shape memory polymers fibers with application to self-healing smart materials. J. Appl. Mech. 80 (1), 1–15, paper number: 011014.
62. Shojaei, A., Voyiadjis, G.Z., Tan, P.J., 2013. Viscoplastic constitutive theory for brittle to ductile damage in polycrystalline materials under dynamic loading. Int. J. Plast. 48, 125–151.
63. Simo, J.C., Hughes, T.J.R., 1997. Computational Inelasticity. Springer, New York.
64. Simo, J.C., Ortiz, M., 1985. A unified approach to finite deformation elastoplastic analysis based on the use of hyperelastic constitutive equations. Comput. Method. Appl. M. 49 (2), 221–245.
65. Sluys, L.J., de Borst, R., 1994. Dispersive properties of gradient-dependent and rate-dependent media. Mech. Mater. 18 (2), 131–149.
66. Steinberg, D.J., 1996. Equation of state and strength properties of selected materials. Livermore, CA, Tech. Rep.UCRL-MA-106439, Lawrance Livermore National Labratory.

67. Thompson, M., Willis, J.R., 1991. A reformulation of the equations of anisotropic poroelasticity. J. Appl. Mech. 58, 612–616.
68. Voyiadjis, G.Z., Faghihi, D., 2012. Thermo-mechanical strain gradient plasticity with energetic and dissipative length scales. Int. J. Plast. 30–31, 218–247.
69. Voyiadjis, Z., Kattan, P.I., 2006. Advances in Damage Mechanics. Elsevier, London.
70. Voyiadjis, G.Z., Mozaffari, N., 2013. Nonlocal damage model using the phase field method: theory and applications. Int. J. Solids Struct. 50 (20–21), 3136–3151.
71. Voyiadjis, G.Z., Alsaleh, M.I., Alshibli, K.A., 2005. Evolving internal length scales in plastic strain localization for granular materials. Int. J. Plast. 21 (10), 2000–2024.
72. Voyiadjis, G.Z., Shojaei, A., Li, G., 2011. A thermodynamic consistent damage and healing model for self healing materials. Int. J. Plast. 27 (7), 1025–1044.
73. Voyiadjis, G.Z., Shojaei, A., Li, G., 2012a. A generalized coupled viscoplastic–viscodamage–viscohealing theory for glassy polymers. Int. J. Plast. 28 (1), 21–45.
74. Voyiadjis, G.Z., Shojaei, A., Li, G., Kattan, P., 2012b. Continuum damage-healing mechanics with introduction to new healing variables. Int. J. Damage Mech. 21 (3), 391–414.
75. Voyiadjis, G.Z., Shojaei, A., Li, G., Kattan, P.I., 2012c. A theory of anisotropic healing and damage mechanics of materials. Proc. R. Soc. A: Math. Phys. Eng. Sci. 468 (2137), 163–183.
76. Xu, W., Li, G., 2010. Constitutive modeling of shape memory polymer based self-healing syntactic foam. Int. J. Solids Struct. 47 (9), 1306–1316.

Chapter 2

Numerical Simulation of Complex Fracture Growth during Tight Reservoir Stimulation by Hydraulic Fracturing

Md. Mofazzal Hossain[a] and M.K. Rahman[b]

[a]Australian School of Petroleum, The University of Adelaide, Adelaide, SA-5005, Australia

[b]School of Oil and Gas Engineering, The University of Western Australia, Perth, WA 6009, Australia

ABSTRACT

The success or failure of hydraulic fracturing technology is largely dependent on the design of fracture configurations and optimization of treatments compatible with the *in-situ* conditions in a given reservoir. The petroleum industry continues to face challenges with this technology in the field, such as premature screen-outs, high treating pressures, complexities with multiple fractures propagation, complex fracture propagation from the deviated wellbore, etc. As these challenges persist better understanding of hydraulic fracture behavior for various reservoir conditions is still an important topic for research. Since the mechanism of hydraulic fracture growth involves the rock stress field and fluid flow field, the modeling work of fracture growth requires the treatment of coupled fluid flow and structural deformation phenomena. In this context, this paper briefly, summarizes an existing numerical tool for fracture growth analysis based on coupled fluid flow and structural deformation phenomena. Solid models have been developed to simulate different field conditions and then solved by using this numerical tool. The field conditions include different stress regimes, fracture geometry and fracture and well orientations. Results for different conditions have been presented and discussed to provide guide lines for better planning and design of hydraulic fracturing. The key finding is that if the well orientation and fracture configuration are not compatible with the *in-situ*stresses, complex fracture growth diminishes the likelihood of success and exhibits some of the above mentioned symptoms during treatments in the field.

INTRODUCTION

Stimulation of tight reservoirs by hydraulic fracturing has been established as a very successful technology for improving the petroleum production performance. Recently, this technology has also been extended to various unconventional applications, such as completion of high permeability unconsolidated formations (FracPacking), geothermal energy resources extraction, waste re-injection, produced water re-injection, coal bed methane gas production, etc. The success of fracture stimulation is largely dependent on the size, shape and the propagation behavior of the created hydraulic fracture. A fracture initiated from a deviated wellbore becomes subject to a complex stress state and leads to the development of a complex geometry of the propagated fracture. This results in a fracture of very limited width. The limited width and the tortuous fracture geometry hinder the flow of injected proppant inside the fracture, which results in a low proppant concentration and hence inadequate fracture conductivity. The ultimate consequence is significant reduction of reservoir productivity, making the whole stimulation process unsuccessful. A recent trend is therefore to develop coupled non-planar fracture models and their use for interesting parametric studies (Hossain, 2001, Dong and de Pater, 2001, Dong and de Pater, 2002, Garcia and Teufel, 2005 and Rungamornrat et al., 2005) to understand the complex fracture growth. This paper is a contribution towards this trend. In this study, an attempt has been made to develop insights to explain how the complex fracture geometry grows under different stress and well conditions, and their effects on the injection pressure, using a flow-deformation coupled numerical tool, HYFRANC3D. This tool was developed by the Cornell Fracture Group (http://fac.cfg.cornell.edu; Carter et al., 2000). The tool considers complex fracture geometry, non-linear coupling between equations that characterize fluid flow in the fracture, structural deformation and possible interaction with other types of fractures (e.g. multiple fractures, natural fractures, etc.).

A vast literature exists that reports the progress of basic understanding and optimization of this technology (Settari and Cleary, 1984, Rahim et al., 1995, Mohaghegh et al., 1999, Soliman and Boonen, 2000,Hossain et al., 2000, Rahman et al., 2001, Queipo et al., 2002, Rahman and Joarder, 2006, Boholi and de Pater, 2006 and Rahman et al., 2007) and different fracture propagation models (Mendelsohn, 1984a,Mendelsohn, 1984b, Veatch and Moschovidis, 1986, Settari and Cleary, 1986, Gidley et al., 1989, Valko and Economides, 1995 and Yew, 1997). Over the years, the works related to the development of fracture models advanced rapidly from 2D analytical models, for instance PKN and KGD models, to complex 3D numerical models. Warpinski et al. (1994) includes brief descriptions and a comparison of predictions for a number of simulators, including 2D and pseudo-3D models. All of these fracture propagation models are based on a similar criterion, which assumes that fracture propagation takes place when certain deformation parameters reach a critical value. For example, according to the linear elastic theory, the fracture is assumed to start propagating once the tensile stress at the fracture tips reaches the tensile strength of the rock. The existence of a stress singularity at or near the fracture tip, however, makes the linear elastic fracture mechanics (LEFM) theory adequate to predict the fracture growth behavior accurately. In LEFM, such a stress singularity is addressed using its strength, termed as stress intensity factor.

Since the mechanism behind hydraulically initiated fracture heavily involves a rock stress field and a fluid flow field, the problem requires the treatment of coupled fluid flow and structural deformation phenomena. It is to be noted that a strong non-linearity exists between the solid structure and the moving fluid. Therefore, the mathematical formulation of an overall fracture propagation model requires coupling of a set of complex equations, hence the development of sophisticated numerical tools based on finite element or boundary element methods. Since the propagation of fracture is mainly controlled by the stress singularity at the fracture tip, it is sufficient to consider the problems on the fracture boundary rather then the whole region as considered in finite element methods. Hence, the boundary element method is usually considered to be more suitable to solve for structural responses due to rock mass, *in-situ* stresses and fracturing fluid pressure. On the other hand, the fluid flow equation can be more conveniently solved by a finite element method. Therefore, the overall computation time to solve a fluid pressure driven fracture propagation problem can be minimized significantly by combining these two numerical methods. Based on this principle, a hydraulic fracture propagation simulator (HYFRANC3D) was developed (Carter et al., 2000). This simulator has been used in this study to investigate the complex fracture growths under different conditions.

In the rest of this paper, we briefly introduce the boundary element method as it applies to modeling the coupled hydraulic fracturing problem and explain the numerical solution technique. We examine a number of case studies, review their results and present final conclusions.

BOUNDARY ELEMENT METHOD

The boundary element method (BEM) is one of the most powerful numerical techniques developed in the shadow of the finite difference and finite element methods. The application of BEM is perfectly feasible to two and three-dimensional fracture problems; however, the code requires a multi-domain discretization capability. This capability facilitates the modeling of two fracture surfaces in separate sub-domains (Luchi and Rizzuti, 1987). The stress intensity factors in 3D fracture geometries can be accurately calculated by a standard BEM code. The application of BEM for solving three-dimensional fracture problems is well documented in the literature (Rizzo, 1967, Cruse, 1969, Lachat and Watson, 1976, Brebbia, 1978, Brebbia et al., 1984 and Luchi and Rizzuti, 1987).

The basic boundary integral equation that provides the relationship between the displacement U, and the traction T, at a surface , of a homogeneous isotropic media, or a sub-domain into which the body has been divided, can be written (Brebbia et al., 1984, Luchi and Rizzuti, 1987 and Banerjee, 1994) as:

$$C_{ij}(A_1)U_j(A_1) + \int_\Gamma F_{ij}(A_1,A_2)U_j(A_2)d\Gamma = \int_\Gamma G_{ij}(A_1,A_2)T_j(A_2)d\Gamma \tag{1}$$

where A_1 and A_2 are the points on the boundary subsurface and the boundary surface, respectively. $F_{ij}(A_1,A_2)$, $G_{ij}(A_1,A_2)$ are the functions representing displacements and tractions, respectively, in the j direction at point A_2 corresponding to a unit point load acting in the i direction applied at A_1, and C_{ij} is a coefficient function. These functions can be obtained from the singular solution for given boundary conditions (Banerjee, 1994). The traction T represents a stress normal to a surface, $T = \sigma n$, where n is unit outward normal. Eq. (1) disregards the body force term and applies to media that follow the linear elastic material behavior.

In order to use the boundary integral equation detailed in Eq. (1) for modeling of hydraulic fracturing problems, the fluid pressure term needs to be included in traction, T. Detailed formulations of such problems are described in the following section.

Mathematical Modeling of Hydraulic Fracture

The modelling of hydraulic fracturing problem is different in the sense that the traction, T, contains the pressure term arising due to fluid flow through the fracture. The displacement, U, is the resultant of that from the fluid flow effect and the structural elastic response. The total displacement gives the fracture aperture that can be expressed as:

$$w = w_0 + wp \tag{2}$$

where w_o is the aperture contribution from the external stress and w_p is that from the fluid pressure. The value of w_p can be expressed as $w_p = \lambda p$, where λ is the influence coefficient. Following the basic boundary integral equation (Eq. (1)) and applying the principle of fluid flow through parallel plates (i.e. $q = -\dfrac{w^3}{12\mu} \text{grad} p$) and mass conservation, the final form of the integral equation for the coupled problem of hydraulic fracturing can be expressed as (Carter et al., 2000):

$$\int_\Omega \delta p \frac{\partial w}{\partial t} d\Omega + \int_\Omega \frac{w^3}{12\mu} (\text{grad } p \cdot \text{grad } \delta p) d\Omega$$

$$+ \int_\Gamma \delta p \; \beta V^{4/3} d\Gamma = Q(t)\delta p(O) \tag{3}$$

in which the bulk fracture is described by a sub-domain Ω in a 3D space F and Γ is the boundary between Ω and the front region of fracture (Fig. 1); p is the pressure inside the fracture; w is the total fracture aperture; q is the flow rate or influx into the fracture; t is the flow time and Q is the source or sink strength of any point O inside a fracture region.

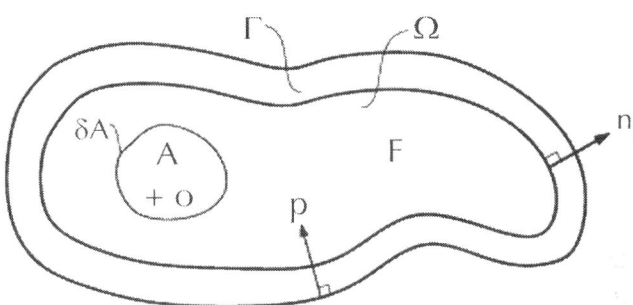

Figure 1: Schematic of the fracture region showing that the fluid enters and leaves a chosen domain A through ∂A in the fracture domain Ω. (after Carter et al., 2000).

The first two integral terms in Eq. (3) represent the usual solution of a hydraulic fracture problem. The third term is a contour integral to incorporate the asymptotic solution for fluid flow. The right hand side of this equation accommodates the boundary condition of a source term at any point (Carter et al., 2000).

It is to be noted that the asymptotic solution of fluid flow involves a strong non-linear coupling between the moving fluid and the solid deformation, particularly in the vicinity of the hydraulically induced fracture tip. The non-linear coupling makes an exact matching singularity between the pressure and the elasticity equations (SCR Geomechanics Group, 1993 and Desroches et al., 1994). The nature and the strength of such a singularity in the fracture depend on fluid properties, formation properties and the fracture propagation speed. Their coupled formulation led to the development of Linear Elastic Hydraulic Fracturing (LEHF) theory. According to LEHF, the solution of crack tip fields is represented by fracture width $w(\rho)$ and expressed for the Newtonian fluid and impermeable formation as, $w(\rho) = \beta V^{1/3}$ where $\beta = 2(3^{7/6})\left(\dfrac{\mu}{E}\right)^{1/3} \rho^{2/3}$, in which ρ is the curvilinear distance measured on the fracture surface between any point and fracture front and μ is the fracturing fluid viscosity and V is fracture propagation speed.

The HYFRANC3D simulator works based on the coupled Eq. (3), where the structural responses are solved by BEM and the fluid flow part is solved by FEM. The detailed finite element solution and nodal discretization of Eq. (3) have been presented by Carter et al. (2000).

NUMERICAL SOLUTION

Since the problem is both time and space dependent, the problem can be solved either by searching a geometry of the propagated fracture for a given injection period, or by searching the injection period for a given geometry of the propagated fracture. The first approach requires more computational time than the second approach and hence the first approach has not been incorporated in HYFRANC3D. As the fracture initiation mechanism is not incorporated in HYFRANC3D, the second approach requires an assumed initial fracture as the starting point which then is modified by subsequent analyses incorporating mixed mode fracture propagation theory. The solution process consists of the following operations:

- Create the boundary element mesh on an assumed fracture geometry (e.g. penny shaped crack) for a given time stage
- Apply the relevant pressure boundary condition to the wellbore (if any)
- Solve for elasticity by boundary element method (BEM)
- Solve for the coupled hydraulic fracture problem, making use of the previous solution (obtained for the $(n-1)$th time stage)
- Allow next stage of fracture propagation
- Repeat steps 1–5 for the new fracture

Following the above solution process, numerical analysis is performed using BEM based code for 3D Hydraulic Fracture Analysis called HYFRANC3D. As mentioned earlier, HYFRANC3D combines the boundary element analysis for structural response with the finite element analysis for fluid flow. The boundary element solution is provided by an external program called BES (see http://www.cfg.cornell.edu). The overall solution methodology consists of the following steps:

- Extend the fracture up to an arbitrary distance based on given boundary condition and discretize the extended fracture,
- Solve the elastic structural deformation problem using BES,
- Solve the fluid flow equation in an iterative fashion until the solution converges to an equilibrium fluid pressure.

The solution process within this loop is complex. To initiate the iterative fracture propagation process, an initial fracture configuration is necessary that satisfies the fracture propagation criterion. This initial fracture configuration can be estimated approximately using analytical linear elastic fracture mechanics principles, or by a number of trials. Once the first step of fracture propagation is complete using the initial configuration, this new solution can be used as the starting point for the next step of propagation. This process continues in subsequent steps of propagation. The iterative solution proceeds at two levels for given initial values of fracture opening, pressure, and fracture front speed. First, for a given time step, the fluid flow equation is solved iteratively using reasonable tolerance on the fracture opening to judge whether the solution has converged. Once this is satisfied, the global mass-balance and the fracture tip speed are checked. The total volume of injected fluid minus the leaked-off volume should equal the fracture volume. Ignoring the fluid lag effect at the tip, the fluid speed at the fracture front is assumed to be equal to the fracture front speed. The consequences of fluid lag effect in 3D hydraulic fractures are well reported by Advani et al. (1997). However, SCR Geomechanics Group (1993) and Carter et al. (2000) have argued, based on their numerical results, that the fluid lag effect on hydraulic fracture width and pressure is negligible.

Both the equations for fluid flow and total volume balance express the satisfaction of global mass-balance. If these two equations are not satisfied, the time step is adjusted and the fluid flow equation is solved iteratively again. This process continues until the solution has converged or the number of global iterations has exceeded the user supplied maximum value. In case of permeable formations, an additional set of iterations for the first stage is considered to ramp up the leak-off from the impermeable to the final permeable solution.

For each stage, the fracture is allowed to propagate step-wise as limited extensions of the previous fracture as shown in Fig. 2.

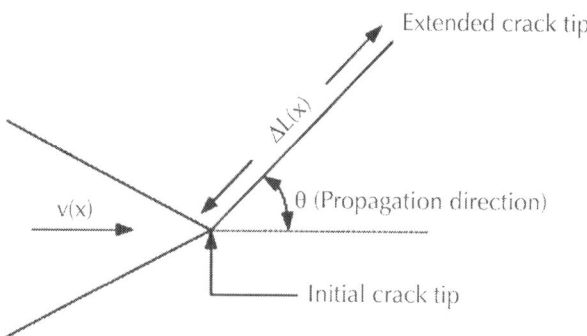

Figure 2: Schematic of fracture propagation process.

The extension is estimated based on the assumption that the local extension of the fracture ΔL(x) at a point x along the fracture front is proportional to the speed of the fluid at that point v(x). The local extension is scaled by a maximum propagation length L_0, specified by the user as:

$$\Delta L(x) = \frac{v(x)}{\max(v)} L_0. \tag{4}$$

The ideal maximum propagation length proves to be in the order of 10% of the fracture penetration into the formation. The ratio of speed and propagation length is later used to determine the initial time step between two subsequent models, as $\Delta L = v\Delta t$. The direction of fracture propagation, θ(x), at each point is determined according to the maximum circumferential stress criterion (Desroches and Carter, 1996 and Rahman et al., 2000) for mixed mode fracture propagation and evaluated in a plane normal to the fracture front at each point and given by:

$$\cos\frac{\theta(x)}{2}\left(K_I\cos^2\frac{\theta(x)}{2} - \frac{3}{2}K_{II}\sin\theta(x)\right) \geq K_{IC} \tag{5}$$

where K_I, K_{II}, K_{IC} are the stress intensity factors for mode I (opening), mode II (shearing) and fracture toughness respectively. K_I and K_{II} are estimated using the displacement correlation techniques (Ingraffea and Manu, 1980, Ingraffea, 1987 and Luchi and Rizzuti, 1987) as functions of the nodal displacements. The existence of mode III (tearing) is not very prevalent in hydraulic fractures emanating from typical well configurations subject to *in-situ* reservoir stresses. Therefore, this mode has not been considered in this study.

CASE STUDIES

Numerical simulations of a number of cases have been performed to investigate the behavior of the propagated fracture geometry from the wellbore under different stress regimes. The effects of perforation orientation and multiple fractures are also investigated.

It is assumed that the fluid is Newtonian and there is no leak-off of fluid to the formation. Both of these assumptions have primary effects on the fracture width and hence effects on fracture volume and injection pressure. However, it is strongly believed that modeling of non-Newtonian fluid behavior and fluid leak-off will not change the general trends of findings in this study.

Model Configuration

A penny shaped fracture of 4 cm radius is considered to be initiated hydraulically from the center of a wellbore of 40 cm diameter. The wellbore is assumed to be located at the center of a block of 200 cm × 200 cm × 200 cm dimensions (Fig. 3). Since the model is considered to be equilibrium in a 3D stress system, analyses of half the block is sufficient due to symmetric boundary conditions. The applied far-field stresses are shown in Fig. 3, with the borehole axis aligned with one of the far-field stress directions. The formation is assumed to be linearly elastic with Young›s modulus 10 GPa, Poison›s ratio 0.25, and fracture toughness 5 MPa$\sqrt{m}$.

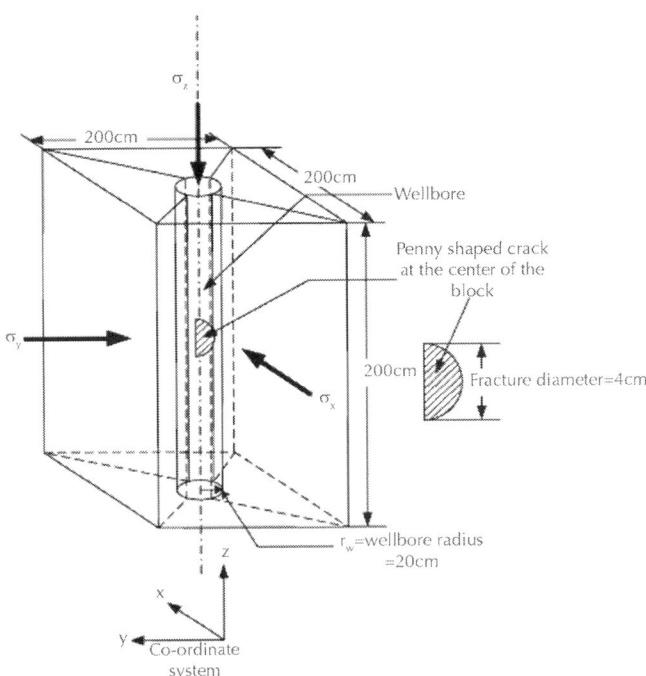

Figure 3: Schematic of the model showing the boundary conditions and necessary dimensions.

Fig. 4 shows the boundary element (meshed) model and applied boundary conditions for a non-perforated well. The far-field *in-situ* reservoir stresses: maximum horizontal stress σ_H; minimum horizontal stress σ_h; and vertical stress σ_v correspond to the wellbore local stresses σ_x, σ_y and σ_z (Fig. 3), respectively. Making appropriate correspondence between the *in-situ* stresses and the local stresses, the actual well orientation can be simulated. For example, Fig. 4 can be considered as a vertical wellbore, when the wellbore axis is parallel to the σ_v direction (i.e. $\sigma_z = \sigma_v$). Similarly, the wellbore is horizontal along σ_h when σ_z is set to σ_H. This technique has been used to avoid repetition of modeling of the geometry.

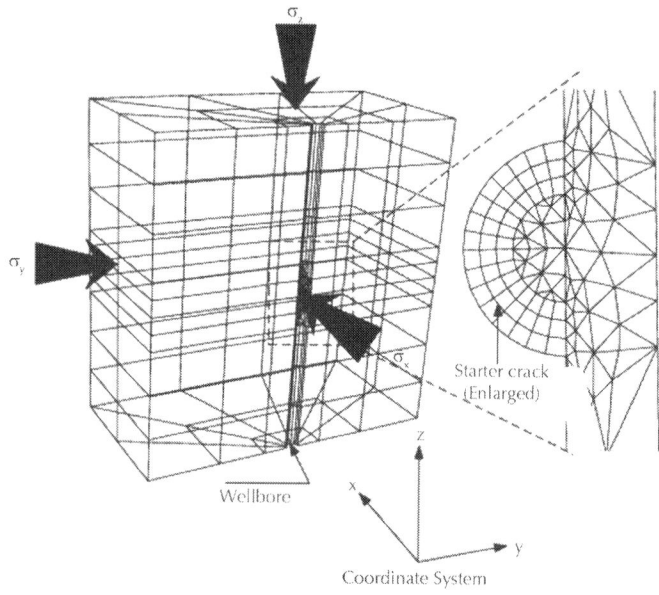

Figure 4: Illustration of boundary element model showing half of the model to represent a wellbore whose axis is parallel to z stress direction with a starter fracture at the center of the block (enlarged view) aligned in the x stress direction.

The fracture propagation study has been carried out into two categories:

- (the structural response with constant fracture pressure
- the coupled fluid and structural response with constant injection rate

The objective of the first category is to understand the behaviour of the fracture geometry when a fracture propagates along preferred and non-preferred directions. The assumption of constant pressure allows the rapid growth of propagated fracture geometry with minimum computational effort. On the other hand, the objective of the second category is to understand the behaviour of the fracture pressure when a fracture propagates along preferred and non-preferred directions.

RESULTS AND DISCUSSION

The Structural Response For Non-Perforated Wellbore

Numerical analysis was performed for a longitudinal fracture (i.e. the fracture plane is parallel to the wellbore axis) and a transverse fracture (i.e. the fracture plane is perpendicular to the wellbore axis). These fractures were initiated from a non-perforated wellbore subject to three different stress regimes (normal faulting, $\sigma_v > \sigma_H > \sigma_h$; reverse faulting, $\sigma_H > \sigma_h > \sigma_v$ and strike-slip, $\sigma_H > \sigma_v > \sigma_h$). The geometric and mechanical properties and the applied stresses are shown in Table 1.

Table 1: Geometric dimensions and rock properties considered for the analysis

Geometric dimensions		Mechanical properties	
Block dimension	200 cm × 200 cm × 200 cm	Elastic modulus (E)	10 Gpa
Fracture shape	Penny shaped	Poisson's ratio	0.25
Fracture radius (half length)	4 cm	Leak-off coefficient	0.0
Fracture length	8 cm	Fracture toughness	5.0 MPa√m
Wellbore diameter	20 cm		
	Applied stresses		
Stress regime	σ_v	σ_H	σ_h
Normal faulting	80 MPa	70 MPa	60 MPa
Reverse faulting	60 MPa	80 MPa	70 MPa
Strike-slip	70 MPa	80 MPa	60 MPa
	Applied constant fluid pressure (MPa)		
Fracture type	Normal faulting	Reverse faulting	Strike-slip faulting
Longitudinal fracture	60 MPa	70 MPa	60 MPa
Transverse fracture	80 MPa	60 MPa	60 MPa

A constant pressure equal to the wellbore pressure is applied inside the fracture; the magnitude of this pressure is set equal to the fracture closure stress. The actual closure stress depends on the fracture orientation and the stress regime considered for a particular case. For example, in the case of longitudinal fractures in the σ_H direction in normal faulting stress regimes, the magnitude of closure stress is taken equal to σ_h.

From an iterative solution of fracture propagation, the stress intensity factors, K_I and K_{II}, at the different lengths of propagation, are obtained. The stress intensity factors for longitudinal and transverse fractures emanating from the vertical wellbore under normal faulting, strike-slip and reverse faulting stress regimes are plotted in Fig. 5, Fig. 6 and Fig. 7, respectively.

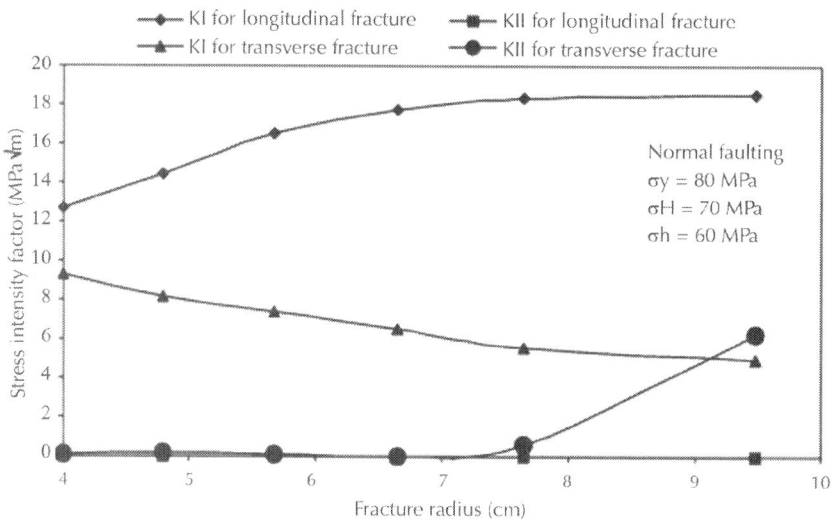

Figure 5: Stress intensity factors, K_I and K_{II} at different propagated fracture radii of longitudinal and transverse fractures for a non-perforated vertical wellbore in a normal faulting stress regime.

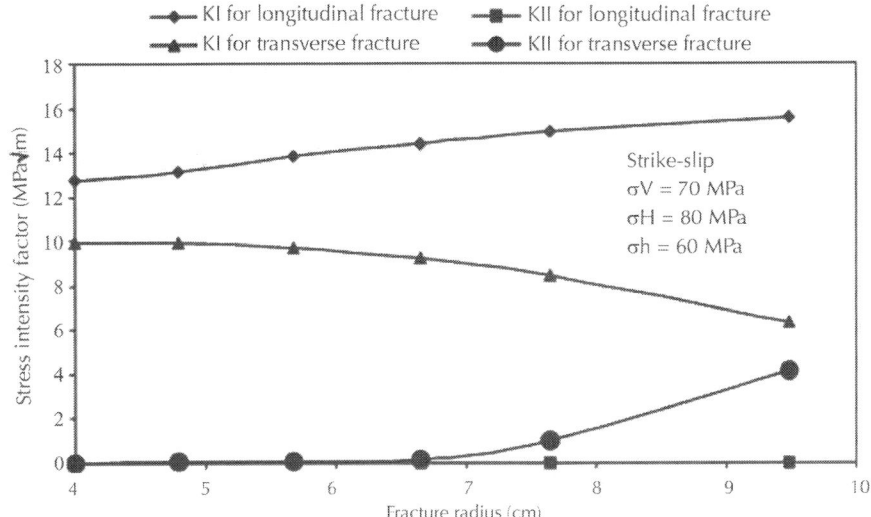

Figure 6: Stress intensity factors, K_I and K_{II} at different propagated fracture radii of longitudinal and transverse fractures for a non-perforated vertical wellbore in a strike-slip stress regime.

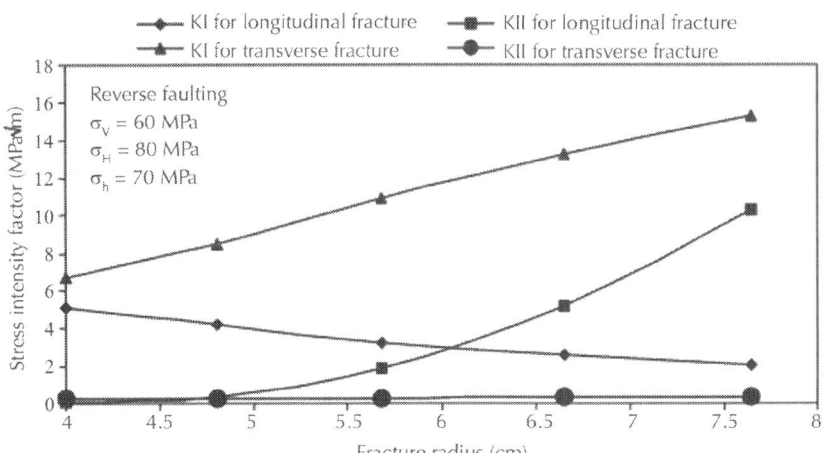

Figure 7: Stress intensity factors, K_I and K_{II} at different propagated fracture radii of longitudinal and transverse fractures for a non-perforated vertical wellbore in a reverse faulting stress regime.

It can be observed from Fig. 5 and Fig. 6 that the stress intensity factor K_I increases for longitudinal fractures from the vertical wellbore in normal faulting and strike-slip stress regimes whereas the value of K_{II} is very negligible and remains constant during fracture propagation. This indicates that the opening mode dominates the fracture propagation process, which results in a planar fracture. This conforms to the general consensus that when a fracture initiates with appropriate orientation in the preferred direction, which is a longitudinal fracture in the maximum horizontal stress direction in a vertical wellbore under normal faulting and strike-slip stress regimes, it propagates in its own plane.

Reverse trends for K_I and K_{II} are observed in Fig. 5 and Fig. 6 for transverse fractures emanating from the vertical wellbore subject to the same stress regimes. This is because the transverse fracture from the vertical wellbore in the normal faulting and strike-slip stress regimes is not favorable for propagation. However, a perfectly transverse fracture should propagate in plane remaining perfectly perpendicular to σ_v though requiring high pressure to overcome high σ_v. This is because the shear component, and hence K_{II}, is zero for such a condition. In reality, a perfect transverse fracture (in other words a perfect planar fracture perpendicular to any stress direction) is not possible in highly heterogeneous reservoir rock. As soon as the fracture goes slightly out of plane, the shear component starts developing to reorient the fracture further towards the preferred direction for fracture propagation with minimum resistance. To mimic this fracture behavior in heterogeneous rock, the transverse fracture in the 2nd step of propagation was made very slightly out of plane. In these cases, K_I decreases and the corresponding value of K_{II} increases as the fracture propagates further. Apparently, this is in contrast with the study of Ingraffea (1987) in which it is stated that a fracture finding itself under substantial mode II loading does not long remain in high K_{II}/K_I domain of interaction, rather it quickly changes its trajectory to minimize or eliminate the K_{II} components. The authors believe that Ingraffea's conclusion is correct when the fracture is already deviated by a certain degree with the preferred direction of fracture propagation. In other words, the fracture has already reached the substantial mode II loading.

In such a case, as soon as the fracture starts propagating, it starts aligning itself with the preferred direction through turning only and consequently, the value of K_{II} starts decreasing. On the contrary, the authors created transverse fracture is absolutely in the opposite phase of the preferred direction and plane, and has not reached the substantial mode II loading state. The alignment of this fracture with the preferred direction and plane requires a total of 90° turning/twisting, which requires a substantial increase of the K_{II} component initially. Once the critical mode II loading state is past, the stress intensity factors K_I and K_{II} would behave during further propagation in the way as Ingraffea explained. However, driving the fracture beyond the critical phase may require a significant increase in pressure inside the fracture. To extend the current investigation up to that phase was not possible due to the requirement of an impracticably long computational time. Only the initial turning and twisting trend of the fracture is shown in Fig. 8. In conclusion, the gradual increase in K_{II} in Fig. 5 and Fig. 6 for the vertical fractures can be justified without any contradiction to Ingraffea's statement, which was correct in the context described by Ingraffea (1987).

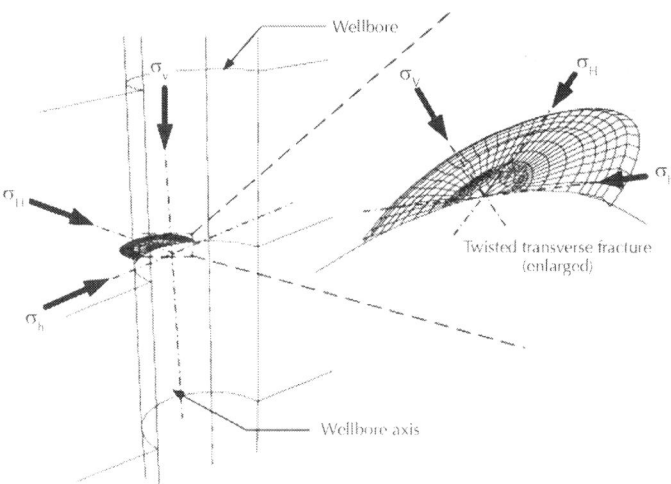

Figure 8: Illustration of twisted transverse fracture emanating from the vertical wellbore in normal faulting stress regime.

On the other hand, in a reverse faulting stress regime the minimum principal stress direction lies in the vertical direction. Therefore, a transverse fracture in such a case should behave similar to the longitudinal fracture in normal and strike-slip stress regimes, and a longitudinal fracture in this case similar to the transverse fracture, for very similar reasons presented in the foregoing paragraph. This common sense expectation is perfectly supported by the trends of stress intensity factors K_I and K_{II} in Fig. 7 and by the geometry of propagated fractures in Fig. 9.

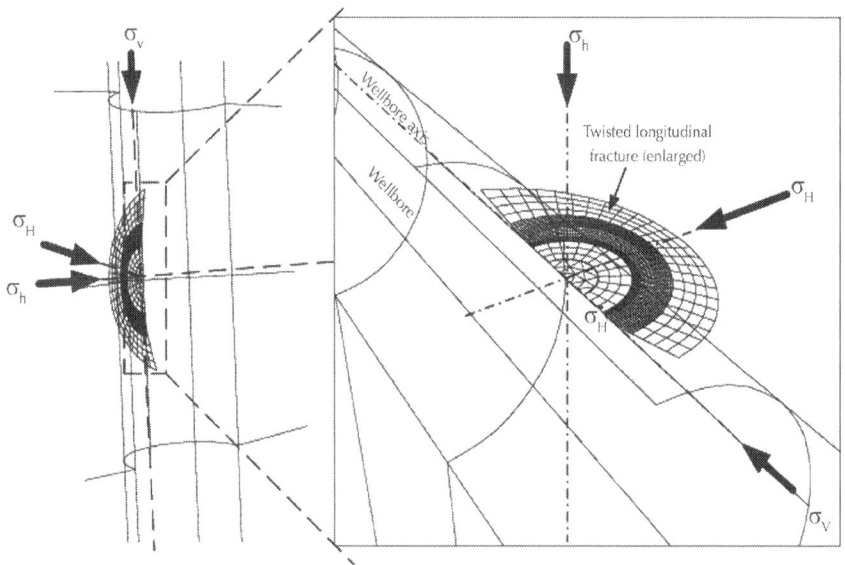

Figure 9: Illustration of twisted longitudinal fracture emanating from the vertical wellbore in reverse faulting stress regime.

In elemental sense, the increase or decrease of K_I and K_{II} depends on the relative magnitudes of nodal deformations along and perpendicular to the fracture around the fracture tip. These deformations, in turn, are complex functions of the current fracture geometry, directions and magnitudes of the applied stresses with respect to the fracture direction and the fluid pressure inside the fracture. Further details can be found in the paper by Ingraffea and Manu (1980). The key finding of this study is, however, if the fracture is already along the preferred direction and the constant fracture pressure is maintained, the value of K_I increases. This indicates that the propagation of this fracture is possible with gradually decreasing fracture pressure. On the other hand, if the fracture is along the non-preferred direction and the constant fracture pressure is maintained, the value of K_{II} increases. In combination with Eq. (5), this indicates that the further propagation of this fracture at some stage may not be possible if the fracture pressure is not increased. This is the direct consequence of out of plane fracture growth.

Structural Response for Perforated Wellbore

This model has the same dimensions as the non-perforated wellbore model. In addition, a perforation has been included from which the fracture has initiated. The perforation tunnel is assumed to be a cone having a face diameter of 1.3 cm, end diameter of 0.65 cm and a length of 18 cm, at the center of the block. A starter fracture of radius 2.5 cm is assumed on each side of the perforation.

The meshed model with a perforation aligned with the maximum horizontal stress is shown in Fig. 10. This model was used to predict the possible location of tension at the interface of the wellbore wall and the perforation from which a fracture is likely to initiate for given stresses and fluid pressure.

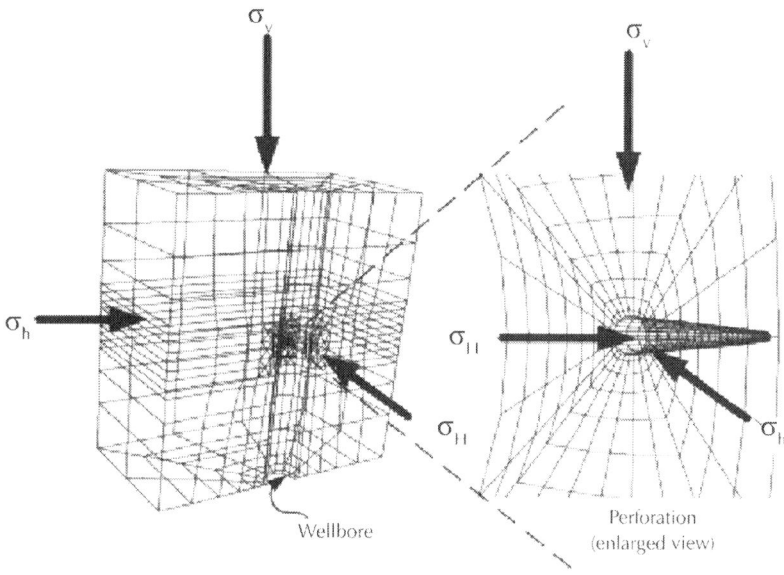

Figure 10: Boundary element model of perforated vertical wellbore, where the perforation is aligned with the direction of σ_H.

The wellbore is vertically situated in a normal faulting stress regime having the stress magnitude: $\sigma_x = \sigma_H = 70$ MPa, $\sigma_y = \sigma_h = 60$ MPa and $\sigma_z = \sigma_v = 80$ MPa. The wellbore pressure is gradually increased until a tensile stress develops in the formation.

Three potential locations are investigated for tensile stress from where the fracture is more likely to initiate and propagate. Location-1 is designated by 0°, which is the mid-depth of the perforation. Location-2 is at 45° up from Location-1 and Location-3 is a further 45° up, which is at the top of the perforation. The stress state at 180° off-phase from each location is identical. The minimum principal stress distribution at these three locations along the perforation length is plotted in Fig. 11. It is observed from Fig. 11 that the minimum principal stress is more tensile (−ve is tension and +ve is compression) in Location-3 compared with the other two locations. This indicates that the probability of initiating the fracture is more likely from Location-3 than the two other locations and will initiate at the base of the perforation. This is highlighted in Fig. 12, where the minimum tangential stress along the perforation face is found at the top or bottom of the perforation circumference (in Fig. 12, when tangential distance is 90°, i.e. Location-3). This simple study indicates that the longitudinal fracture is likely to happen from a vertical wellbore in a normal faulting stress regime even after the wellbore is perforated. However, the possibility of multiple fractures from different locations can not be dismissed as all three locations are in tension. Their propagation behavior is therefore studied further.

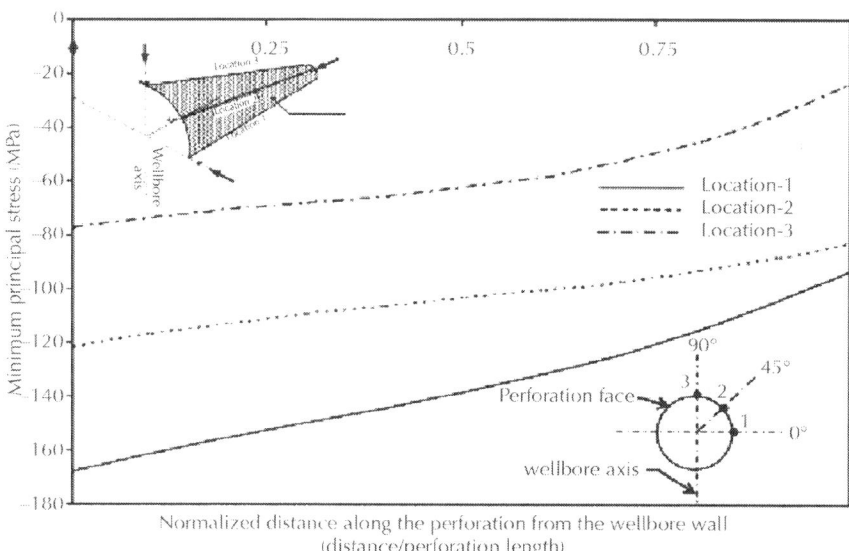

Figure 11: Maximum principal stress distribution along the perforation from the wellbore wall.

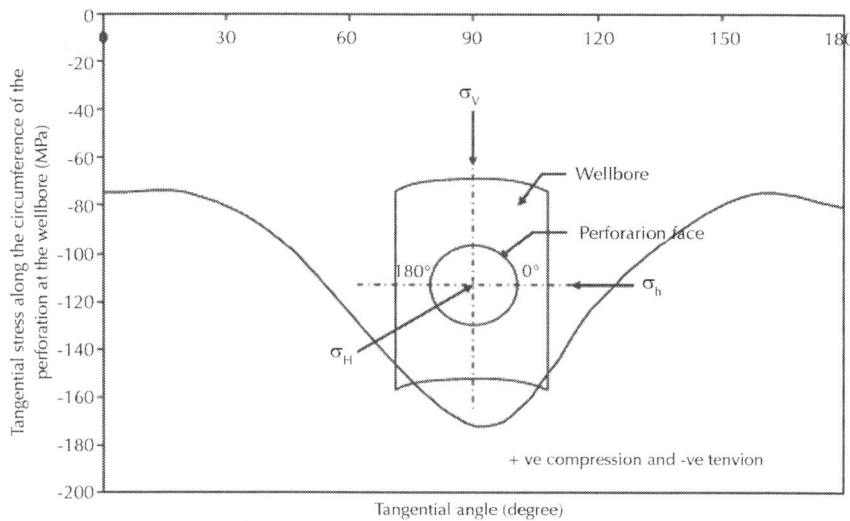

Figure 12: Tangential stress distribution along the base of perforation on the wellbore wall.

Fractured Perforated Wellbore

To investigate the effect of fracture orientation with respect to *in-situ* stress direction with the three different stress regimes, a longitudinal and a transverse fracture of 2.5 cm radius are created as starter fractures on both sides of a perforation face at the vertical wellbore (Fig. 13). The applied stresses, pressures and mechanical properties are the same as used for the non-perforated wellbore as given in Table 1. The perforation is considered aligned with the direction of σ_H for all the three stress regimes.

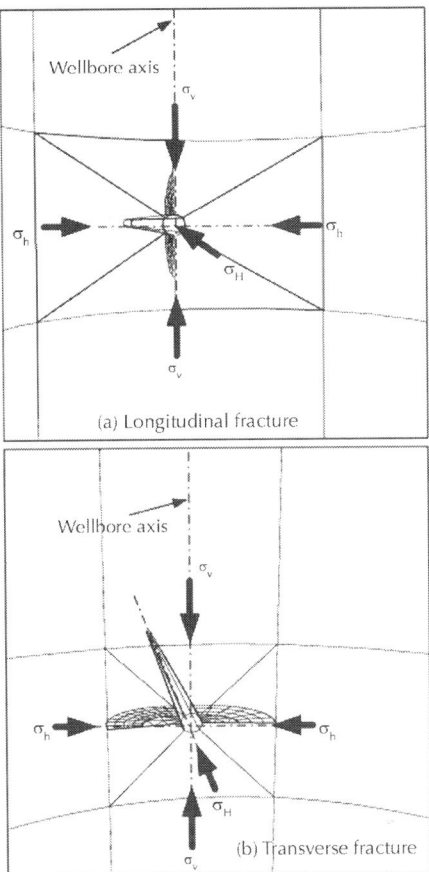

Figure 13: Illustration of longitudinal and transverse fractures emanating from the perforated vertical wellbore, when the perforation is oriented in the direction of $_H$.

Fig. 14 presents the stress intensity factors, K_I and K_{II}, at different fracture propagation radii in the normal faulting stress regime for both longitudinal and transverse fractures. The figure shows the same trends of stress intensity factors with slightly different numerical values, as that with the non-perforated wellbore. Therefore, the fracture propagation behavior is also expected to be similar to that for the non-perforated wellbore. This is also manifested by the turning–twisting trend of a transverse fracture in Fig. 15. Similar results are expected for strike-slip and reverse faulting stress regimes and therefore they are not studied further with perforated wellbores.

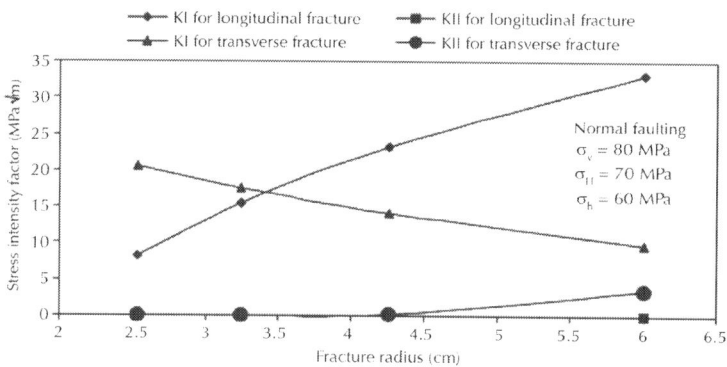

Figure 14: Stress intensity factors, K_I and K_{II}, at different propagated fracture radii of longitudinal and transverse fractures for a perforated vertical wellbore in a normal faulting stress regime.

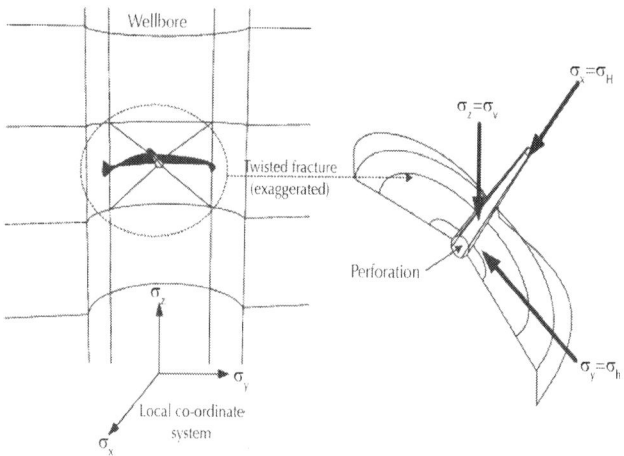

Figure 15: Propagated transverse fracture initiated from the perforation aligning in the $_H$ direction for vertical wellbore in normal faulting stress regimes.

Coupled Fluid and Structural Response

From the above case studies, it is now clearly established that the fracture initiated in the non-preferred direction develops complex geometry through turning and twisting during propagation. This is expected to result in flow constriction and high treating pressure. In order to investigate the nature of pressure response during complex hydraulic fracture propagation, a case of coupled fluid flow and structural response problem has been studied. A horizontal wellbore subject to the normal faulting stress regime is considered to align along the $_h$ direction for this case. A longitudinal fracture is initiated from the preferred as well as non-preferred directions. The effect of multiple fractures on the treating pressure and fracture volume is also investigated.

Fracture in the Preferred vs Non-preferred Direction

Fig. 16 represents the horizontal wellbore oriented along the $_h$ direction. The applied *in-situ* stresses are the stress along the wellbore axis, $_h$ = 44 MPa, the vertical stress, $_v$ = 77 MPa and the other horizontal stress, which is orthogonal to the wellbore axis, $_H$ = 56 MPa. The fluid is considered to be a Newtonian one having a viscosity of μ = 100 cp. A constant flow rate of 7.5 cm^3/s is simulated in the fracture mouth. The fluid leak-off is ignored. The mechanical properties of the formation are used as presented in Table 1. A penny shaped longitudinal fracture was initiated along the preferred direction first, i.e. along $_v$. The height (diameter) of the initiated fracture mouth at the wellbore wall was 10 cm, which was then propagated stepwise. Similar operations were performed with an initiated fracture along the non-preferred direction, i.e. along $_H$. Pressures developed at the fracture mouth during fracture propagation and the overall fracture volumes for both cases are plotted in Fig. 17, representing the first case by M1 and the second case by M2. Results in Fig. 17 clearly show that the fracture initiated along the non-preferred direction (M2) requires very high pressure for propagation and the fracture volumes tends to reduce after a small extent of fracture propagation.

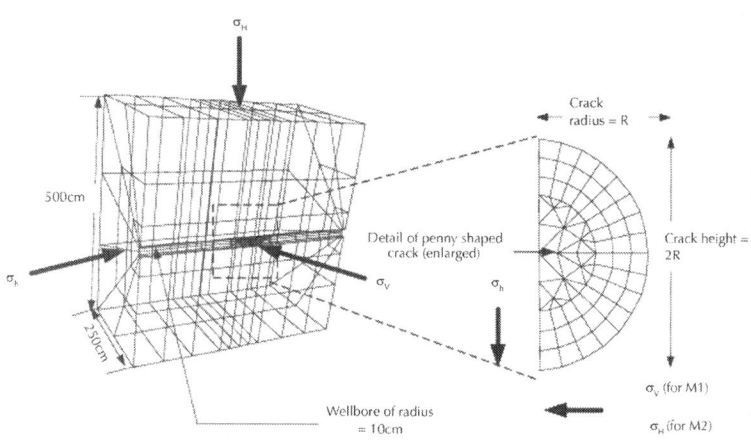

Figure 16: Details of boundary element models of a horizontal well aligned along the minimum horizontal stress (half of block is considered due to symmetry).

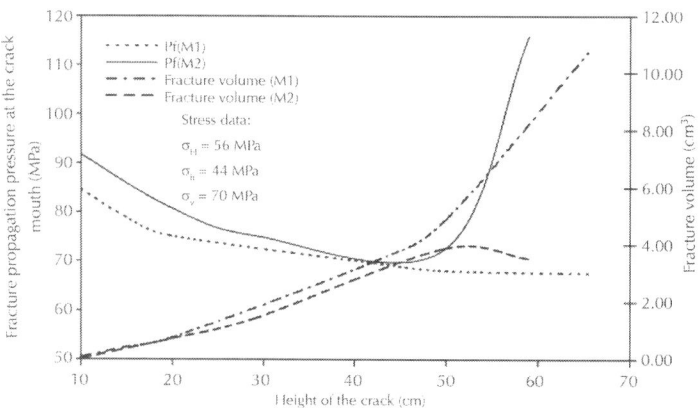

Figure 17: Comparison of propagation pressures and volume of fractures initiated at preferred (M1) and non-preferred (M2) direction.

The visualization of complex fracture growth along the non-preferred direction (Fig. 18) allows one to comprehend the physical phenomena that lead to increasing injection pressure and decreasing fracture volume. Clearly, the fracture has turned and twisted to orient itself with the preferred plane. On the other hand, the fracture initiated along the preferred direction (M1) is more favorable for propagation in terms of both propagation pressure and fracture volume.

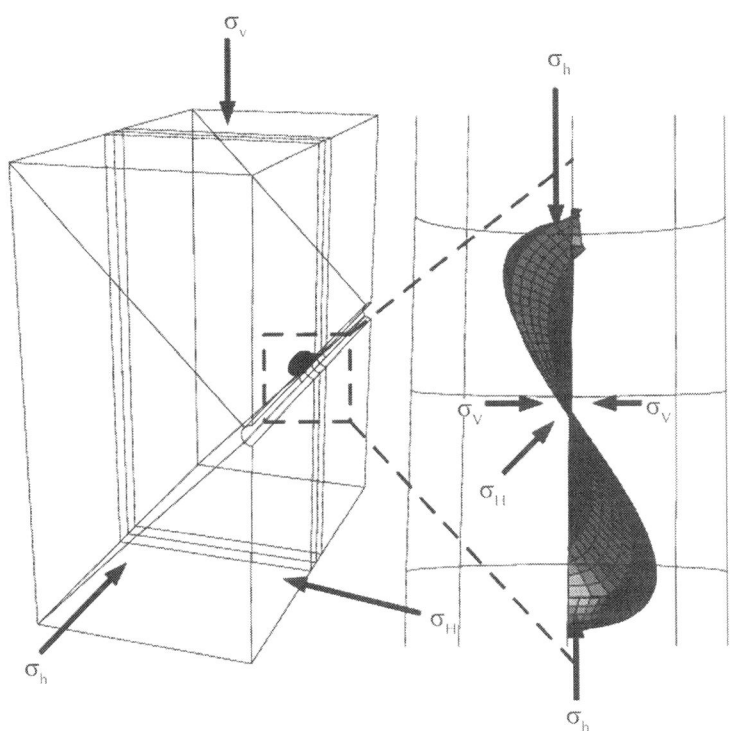

Figure 18: Detail finite element meshing, and tortuous fracture plane during propagation of fracture emanating from the horizontal wellbore in the non-preferred direction obtained from HYFRANC3D analysis.

The pressure response as a function of fracture propagation in this small-scale problem studied has, however, clearly demonstrated that the complex fracture growth in the field may lead to the requirement of a treatment pressure which may be beyond the practical pump capacity. Also the fracture volume, and hence flow conduit, may eventually offer very little predicted production benefits. These two together clearly explain the reasons for screen-outs and non-productive outcomes of many practical fracture treatments.

Single Fracture vs Multiple Fracture

In this case study, single and multiple parallel transverse fractures are modeled. Laboratory tests were conducted by Crosby (1999) for these fracture configurations. The laboratory stress conditions and treatment parameters are summarized in Table 2. The numerical models for the single and multiple fractures and the applied boundary conditions are illustrated in Fig. 19 and Fig. 20 respectively. Note that only half of the block was discretized to take the advantages of symmetric boundary conditions. Fluid leak-off in both models was set to zero again.

Table 2: Summary of laboratory stress conditions and treatment parameters

Test no.	Fracture radius (mm)	Fracture configuration	Fracture test material	Injection rate (Q) (cm^3/s)	Fracture viscosity (μ) (Pa s)	Stresses applied to block (MPa)		
						σ_x	σ_y	σ_z
DC-3	10 of each fracture	Multiple Parallel Fracture	Silica, Cement	0.011[a]	81	7	6	5
DC-4	10	Single completion	Silica, cement	0.024[b]	53	7	6	5

[a] Total injection rate.
[b] Injection rate into each fracture.

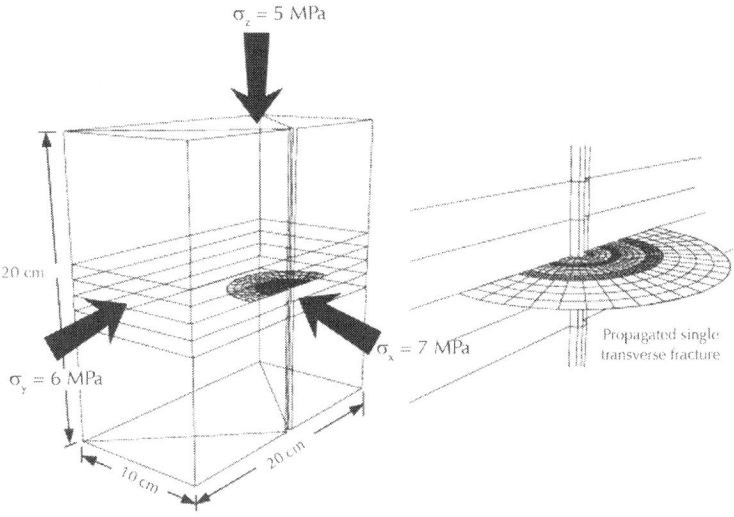

Figure 19: Illustration of single transverse fracture model with boundary conditions analyzed by HYFRANC3D and propagated fracture's configuration.

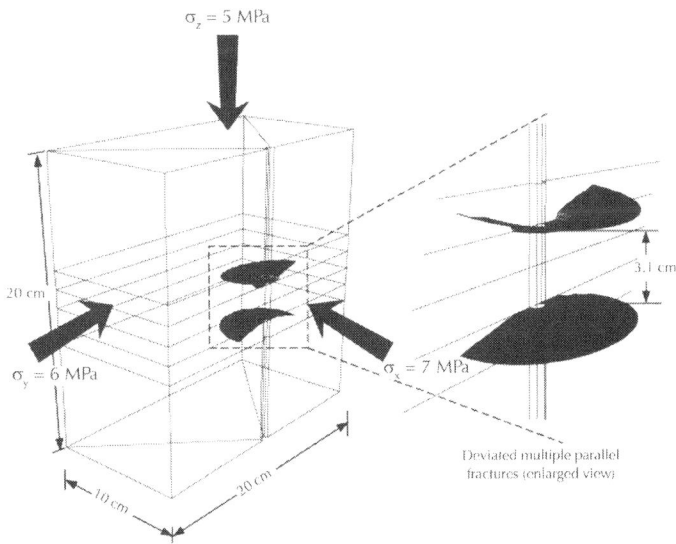

Figure 20: Illustration of multiple parallel fracture model and divergence of propagated fractures.

The stress intensity factors (K_I and K_{II}) at the fracture tip for both single and multiple parallel fractures are plotted in Fig. 21 as functions of the propagated fracture radius. The corresponding fracture propagation pressure and the fracture volume are plotted in Fig. 22.

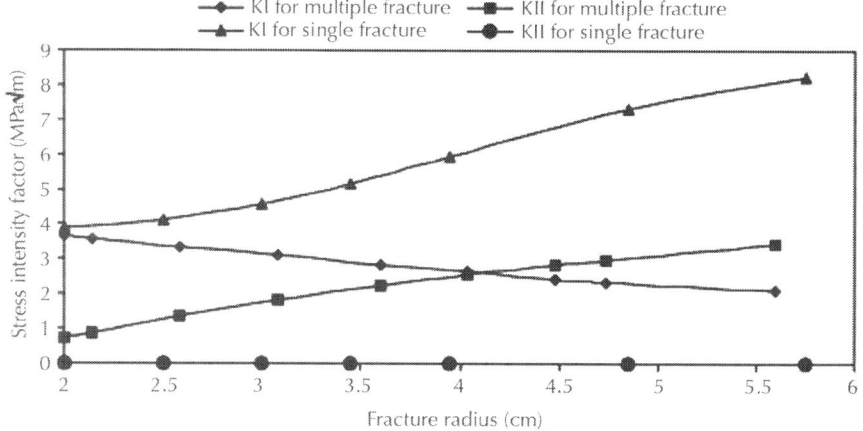

Figure 21: Mode I and mode II stress intensity factors at different propagated fracture radii of single and multiple parallel fractures.

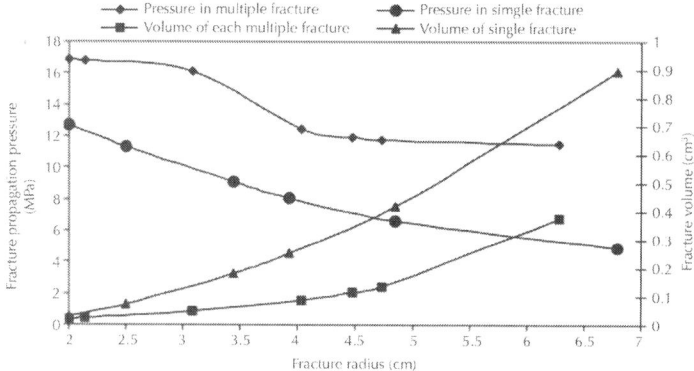

Figure 22: Comparison of propagation pressures and fracture volume of parallel multiple and single fractures.

From Fig. 21, it is observed that the value of K_I for the single fracture considerably increases with the propagated fracture radius keeping the very insignificant K_{II} constant. The multiple parallel fractures have shown the reverse effect, particularly the increasing nature of K_{II}. This indicates turning and twisting of multiple fractures during propagation. This effect has inflicted a divergent nature of propagation of multiple parallel fractures as shown in Fig. 20. As a consequence of this non-planar fracture growth, the injection pressure has increased and the fracture volume has decreased as functions of fracture propagation (Fig. 22). These findings are also supported by the experimental results of Crosby (1999).

CONCLUSIONS

Based on the results and discussions, the following major conclusions can be made.

- It is a very general perception that the fracture initiated in the non-preferred direction and plane turns and twists during propagation and tends to be aligned with the preferred direction and plane. However, if a perfectly planar fracture is perfectly oriented along the non-preferred direction, theoretically the fracture should propagate in plane though may require higher pressure than that for the fracture in the preferred direction. In the field, however, the rock formation is extremely heterogeneous which is more likely to induce out of plane fracture growth. Whenever the fracture propagation is simulated along the non-preferred direction in this study, a slightly out of plane fracture configuration is introduced at the early propagation stage to mimic the field nature. The mixed mode fracture propagation model then convoluted the non-planar fracture growth further, as expected in the non-preferred direction.
- A longitudinal fracture from a vertical wellbore subject to a normal faulting or strike-slip stress regime is favorable for propagation along the maximum horizontal stress direction

without turning and twisting. A transverse fracture is favorable for propagation from such a well if the stress regime is reverse faulting. Fractures initiated otherwise turn and twist during propagation and lead to complex geometry. This behavior remains unchanged even if the wellbore is perforated along the preferred direction. In a number of cases studied, these have been successfully simulated by the numerical tool, HYFRANC3D, used in this study. The tool has rightly facilitated the coupled fluid flow and deformation analysis capability that is crucial to accurately model the fluid driven propagating behavior of hydraulic fractures.

- There is a possibility of initiation of multiple fractures at the base of the perforation although the fracture at a certain location dominates the others. The top as well as the bottom of the perforation base along the maximum horizontal stress is such a likely location for longitudinal fracture initiation from a vertical wellbore subject to a normal faulting stress regime.
- Any perforation which is at off-phase with the preferred direction causes fracture turning and twisting.
- Regardless of the causes of multiple fractures and fracture turning and twisting during propagation, they are the source of high treating pressures and reduced fracture volume. These have been clearly simulated for the small-scale case problems studied in this work. The results, however, have clearly indicated that the complex fracture growth from the non-preferred direction is the most likely reason for premature screen-outs in many fracture treatments in the field. Also the resulting reduced fracture volume is one of the major reasons for reduced productivity.
- It is very important to optimize the well trajectory, perforation direction and fracture configurations for a given stress condition in the field to avoid the treatment failures related to complex fracture growth. Brief guidance includes: (1) oriented perforation along $_H$ direction for vertical wells subject to

normal faulting stress regimes; (2) longitudinal fracturing in vertical wells subject to reverse faulting stress regime is very unlikely to be successful; (3) multiple transverse fracturing of vertical wells subject to reverse faulting stress regimes and horizontal wells along σ_h direction are likely to be successful; however, optimizing the transverse fracture spacing is crucial to avoid non-planar transverse fracture growth which potentially diminishes the predicted treatment benefits.

ACKNOWLEDGMENTS

The authors gratefully acknowledge the contributions of A/Prof. Sheikh Rahman, School of Petroleum Engineering, University of New South Wales and the English editing service of Ms Therese Ellis, School of Oil and Gas Engineering, University of Western Australia. Finally, the two anonymous reviewers deserve thanks for their constructive comments which improved this paper.

REFERENCES

1. Advani, S.H., Lee, T.S., Dean, R.H., Pak, C.K., Avasthi, J.M., 1997. Consequences of fluid lag in three-dimensional hydraulic fractures. Int. J. Numer. Anal. Methods Geomech. 21, 229–240.
2. Banerjee, P.K., 1994. The Boundary Element Methods in Engineering, 2nd edition. Mc-Graw-Hills Book Company, New York, U.S.A.
3. Bohloli, B., de Pater, C.J., 2006. Experimental study on hydraulic fracturing of soft rocks: influence of fluid rheology and confining stress. J. Pet. Sci. Eng. 53, 1–12.
4. Brebbia, C.A., 1978. The Boundary Element Method for Engineers. Pentech Press, London, UK.
5. Brebbia, C.A., Telles, J.C.F., Wrobel, L.C., 1984. Boundary Element Techniques: Theory and Applications in Engineering. Springer- Verlag, New York, U.S.A.

6. Carter, B.J., Desroches, J., Ingraffea, A.R., Wawrzynek, P.A., 2000. Simulating fully 3D hydraulic fracturing. In: Zaman, M., Booker, J., Gioda, G. (Eds.), Modeling in Geomechanics. Wiley Publisher, p. 30.
7. Crosby, D.G., 1999. The initiation and propagation of interaction between, hydraulic fractures from horizontal wellbores, PhD thesis, School of Petroleum Eng., UNSW, Sydney, Australia, July. Cruse, T.A., 1969. Numerical solution in three-dimensional elastostatics. Int. J. Solids Struct. 5, 1259–1274.
8. Desroches, J., Carter, B.J., 1996. Three dimensional modeling of hydraulic fracture. Proc. of 2nd North American Rock Mechanics Symp. (NARMS), Montreal, Quebeck, pp. 995–1002.
9. Desroches, J., Detournay, E., Lenoach, B., Papanastasiou, P., Pearson, J.R.A., Theiercelin, M., Cheng, A., 1994. The crack tip region in hydraulic fracturing. Proc. R. Soc. Lond., A 447, 39–49.
10. Dong, C.Y., de Pater, C.J., 2001. Numerical implementation of displacement discontinuity method and its application in hydraulic fracturing. Comput. Methods Appl. Mech. Eng. 191, 745–760.
11. Dong, C.Y., de Pater, C.J., 2002. Numerical modeling of crack reorientation and link-up. Adv. Eng. Softw. 33, 577–587.
12. Garcia, J.G., Teufel, L.W., 2005. Numerical simulation of fully coupled fluid-flow/geomechanical deformation in hydraulically fractured reservoirs. SPE Paper 94062, 2005 SPE Production and Operations Symposium, Oklahoma, USA. April 17–19.
13. Gidley, J.L., Holditch, S.A., Nierode, D.E., Veatch, R.W., 1989. Recent Advances in Hydraulic Fracturing. Monograph Series, vol. 12. SPE, Richardson, TX.
14. Hossain, M.M., 2001. Reservoir Stimulation by Hydraulic Fracturing: Complexities and Remedies with Reference to Initiation and Propagation of Induced and Natural Fractures, PhD Thesis, The University of New South Wales, Australia.

15. Hossain, M.M., Rahman, M.K., Rahman, S.S., 2000. Hydraulic fracture initiation and propagation: roles of wellbore trajectory, perforation and stress regimes. J. Pet. Sci. Eng. 27, 129–149.
16. Ingraffea, A.R., 1987. Theory of crack initiation and propagation in rock. In: Atkinson, B.K. (Ed.), Fracture Mechanics of Rock. Academic Press Inc., London, pp. 71–110.
17. Ingraffea, A., Manu, C., 1980. Stress-intensity factor computation in three dimensions with quarter-point element. Int. J. Numer. Methods Eng. 15, 1427–1445.
18. Lachat, J.C., Watson, J.O., 1976. Effective numerical treatment of boundary integral equations: a formulation for three-dimensional elastostatics. Int. J. Numer. Methods Eng. 10, 991–1005.
19. Luchi, M.L., Rizzuti, S., 1987. Boundary elements for threedimensional elastic crack analysis. Int. J. Numer. Methods Eng. 24, 2253–2271.
20. Mendelsohn, D.A., 1984a. A review of hydraulic fracture modeling-I: general concepts, 2D models, motivation for 3D modeling. J. Energy Resour. Technol. 106, 369–376.
21. Mendelsohn, D.A., 1984b. A review of hydraulic fracture modeling-II: 3D modeling and vertical growth in layered rock. J. Energy Resour. Technol. 106, 543–553.
22. Mohaghegh, S., Balanb, B., Platon, V., Ameri, S., 1999. Hydraulic fracture design and optimization of gas storage wells. J. Pet. Sci. Eng. 23, 161–171.
23. Queipo, N.V., Verde, A.J., Canelon, J., Pintos, S., 2002. Efficient global optimization for hydraulic fracturing treatment design. J. Pet. Sci. Eng. 35, 151–166.
24. Rahim, Z., Holditch, S.A., Zuber, M.D., Buehring, D., 1995. Evaluation of fracture treatments using a layered reservoir description: field examples. J. Pet. Sci. Eng. 12, 257–267.
25. Rahman, M.K., Joarder, A.H., 2006. Investigating production-induced stress change at fracture tips: implications for a novel hydraulic fracturing technique. J. Pet. Sci. Eng. 51, 185–196.

26. Rahman, M.K, Hossain, M.M., Rahman, S.S., 2000. An analytical method for mixed-mode propagation of pressurized fractures in remotely compressed rocks. Int. J. Fract. 103, 243–258.
27. Rahman, M.M., Rahman, M.K., Rahman, S.S., 2001. An integrated model for multiobjective design optimization of hydraulic fracturing. J. Pet. Sci. Eng. 31, 41–62.
28. Rahman, M.K., Suarez, Y.A., Chen, Z., Rahman, S.S., 2007. Unsuccessfully hydraulic fracturing cases in Australia: investigation into causes of failures and their remedies. J. Pet. Sci. Eng. 57, 70–81.
29. Rizzo, F.J., 1967. An integral equation approach to boundary value problems of classical elastostatics. Q. Appl. Math. 25, 83–95.
30. Rungamornrat, J., Wheeler, M.F., Mear, M.E., 2005. A numerical technique for simulating nonplanar evolution of hydraulic fractures. SPE Paper 96968, 2005 SPE Annual Technical Conference and Exhibition, Dallas, Texas, USA. Oct.9–12.
31. SCR Geomechanics Group, 1993. On the modeling of Near tip processes in hydraulic fractures. Int. J. Rock Mech. Min. Sci. Geomech. Abstr. 30 (3), 1127–1134.
    1. Settari, A., Cleary, M.P., 1984. Three-dimensional simulation of hydraulic fracturing. J. Pet. Technol. 1177–1190.
    2. Settari, A., Cleary, M.P., 1986. Development and testing of a pseudothree- dimensional model of hydraulic fracture geometry. SPE Trans. AIME 281, 449–466.
    3. Soliman, M.Y., Boonen, P., 2000. Rock mechanics and stimulation aspects of horizontal wells. J. Pet. Sci. Eng. 25, 187–204.
    4. Valko, P., Economides, M.J., 1995. Hydraulic Fracture Mechanics. John Wiley and Sons, Chichester, England.
    5. Veatch Jr., R.W., Moschovidis, Z.A., 1986. An overview of recent advances in hydraulic fracturing technology. Soc. Pet. Eng. J. 26, 421–454.

6. Warpinski, N.R., Moschovidis, Z.A., Parker, C.D., Abou-Sayed, I.S., 1994. Comparison study of hydraulic fracturing models—Test case: GRI staged field experiment. SPE Production and Facilities, vol. 3, pp. 7–16. Feb.
7. Yew, C.H., 1997. Mechanics of Hydraulic Fracturing. Gulf Publishing Co., Houston.

Chapter 3

Efficient Optimization Framework for Integrated Placement of Horizontal Wells and Hydraulic Fracture Stages in Unconventional Gas Reservoirs

Xiaodan Ma, Eduardo Gildin, and Tatyana Plaksina [1]

Department of Petroleum Engineering, Texas A&M University, USA

ABSTRACT

Rapid advances in horizontal well drilling and hydraulic fracturing have made these technologies standard development strategies in unconventional gas reservoirs. Further improvements in these practices by means of numerical optimization of wellbore locations and hydraulic fracture (HF) stages spacing can enhance shale gas reserves and increase revenue from the unconventional projects. In order to solve these two challenges simultaneously as an integrated optimization problem, an automated framework for placement of horizontal wellbores and HF stages is developed and tested in this paper. Coupled with expert knowledge and engineering judgment, this workflow allows to produce unconventional assets economically.

This paper presents specifics of our novel optimization framework that improves the design and placement of HF stages in shale gas reservoirs and increases production and the net present value (NPV) of the projects by judicious application of numerical optimization algorithms. In particular, we test several gradient-based and gradient-free methods, namely, simultaneous perturbation stochastic approximation (SPSA), Genetic Algorithm (GA), and covariance matrix adaptation evolution strategy (CMA-ES). Application of these optimization strategies to a suite of test cases illustrates that it is not necessary to assume even spacing between HF stages because the algorithms have a capability to optimize HF stages spacing in homogeneous and heterogeneous geologic systems.

INTRODUCTION

Unconventional resources, such as tight gas sands and shale gas reservoirs, are reshaping the energy supply structure in the United States and are being established as the main cleaner energy sources in the twenty first century (Curtis, 2002 and Jenkins and Boyer, 2008). Economic production of natural gas from shale formations requires favorable petrophysical properties and good well completion potential. Successful completion design depends heavily on a given well location. Therefore, optimal choice of well placement location as well as the number and spacing of hydraulic fractures (HF) stages is critical for meeting commercial production goals.

Hydraulic fracturing operations in gas-rich shale reservoirs tend to be complex and capital consuming (Holditch, 2007). This new technique has been changing the energy future worldwide (Energy Information Administration, 2010). In order to bring the costs down and facilitate the best development practices, reservoir and production engineers might want to utilize an automated approach to wellbore and HF stages placement that can effectively search the wide domain of the objective function for the optimal solution. Benefits of the automated framework are hard to overestimate. Although better petrophysical characterization of shale formations and the engineers' judgment can reduce the search space significantly, optimization algorithms are still the most rigorous strategies for obtaining specific values for desired control variables in a systematic fashion (Cipolla, 2009). There are several optimization works that designed evenly or uniform HF spacing frameworks in shale gas reservoir with given fixed HF locations (Holt, 2011 and Yu and Sepehrnoori, 2013). We propose the numerical optimization workflow that can be used in combination with the expert knowledge to enhance gas reserves and increase revenues from shale gas projects.

Below we formulate the optimization problem of horizontal well placement and spacing of HF stages mathematically, and introduce the details of the proposed framework. The framework is a hierarchical optimization problem with two levels. On the upper level our workflow searches for the best wellbore location or locations (in case of more than one well). Once such location is calculated, it is fixed and passed to the lower level of the workflow. On this level one of the chosen algorithms computes the optimal number, locations and spacing of HF stages by varying the control vector and evaluating the net present value (NPV) objective function.

To solve the discrete optimization problem described above, we employ and compare several algorithms: Simultaneous perturbation stochastic approximation (SPSA), Genetic Algorithm (GA), and covariance matrix adaptation evolution strategy (CMA-ES). All three algorithms are implemented and used to investigate the discrete hierarchical problem of wellbore and HF stages placement. Results of the synthetic test runs reveal advantages and shortcomings of each algorithm and demonstrate clear benefits of our systematic approach to shale gas development.

BACKGROUND AND METHODOLOGY

This section discusses most general mathematical statement of our optimization problem and defines specifics applicable to HF and well placement optimization. The long-term objective function is described with key economic parameters and production variables. The section ends with rigorous introduction to the three optimization algorithms that lie in the heart of our framework.

Objective Function

To define the optimization problem of wellbore and HF placement, we first formulate the objective function J that allows comparing results of all test cases on a common basis. One of the most popular objective functions in oil and gas industry is NPV. In a general mathematical framework, the optimization problem can be stated as follows: find the optimal locations of HF stages u such that

$$u^* = \arg\max_{u \in U} J(u), \tag{1}$$

where $J(u)$ is the NPV objective function with key economic parameters. The search space for the optimal solution u^* is the number and locations of possible HF stages. In our numerical simulation framework u is also labeled as a control vector of integer numbers.

The NPV function is a complex mathematical expression that describes long-term project objectives. It contains terms accounting for the cost of each HF stage and the number of HF stages. In addition, the objective function J includes gas production and water disposal rates as well as drilling and operational costs. All production scenarios are tested on the twenty years production period with some reasonable approximation of the discount rate. The NPV function that optimizes locations and number of HF stages of equal half-lengths as well as well locations has the following form:

$$J = \sum_{k=1}^{K} \left[\sum_{j=1}^{N_{prod}} \frac{\left(Q_{g,j}^k \cdot r_g - r_w - Q_j\right) \cdot \Delta t^k}{(1+b)^{t^k/365}} \right] - \sum_{j=1}^{N_{prod}} (C_w \cdot j + H_j C_f + H_j C_p) \quad (2)$$

In this expression, the first summation term stands for the discounted revenue from the well operations and the second term accounts for drilling and fracturing costs (Holt, 2011). Each parameter of the function J is defined as follows: K is the total number of simulation time steps, k is the time index, t^k[year] is the length of time period, and b is the discount rate [%/100/year]. N_{prod} is the number of production wells, $Q_{g,j}^k$ is gas production rate for a producer j [Mscf/day] at year k, and r_g is constant gas price [\$/Mscf]. In order to describe project's operational and capital expenses, we use Q_j [\$/day] as the operating cost of the well j, C_w [\$] as the base cost of drilling a horizontal well, C_f [\$] as the hydraulic fracturing cost per stage, H_j as the number of HF stages along the well j, and, finally, C_p [\$] as the drilling penetration cost of a gridblock. Table 1 provides specific values for the main parameters of the objective function J.

Table 1: Parameter values for the NPV function. (Schweitzer and Bilgesu, 2009 and Bruner and Smosna, 2011)

Property	Unit	Value
Gas price (at the wellhead)	$/ft^3	3.2
Cost of water disposal	$/bbl	1.0
Discount rate	%/100	13
Base cost for drilling per well	$	2.00E + 06
Penetration cost per gridblock	$	6.00E + 03
Fracturing cost per HF stage	$	1.30E + 05
Operating cost per well	$/day	60

Workflow for Production Design Optimization

Now that we defined the objective function J for our discrete optimization problem, we propose the optimization workflow that in combination with the expert knowledge can enhance gas reserves and increase revenues from shale gas development. Most proposed solutions for the HFs placement problem simply assume all HF stages are spaced with uniform distance between each other. We develop the optimization workflow that allows for non-even HF stages spacing, because our optimization algorithms remove this constraint and automatically selects both the number and locations of the stages. Before we present the key findings and recommendations of our study, we explain the conceptual design of our novel optimization workflow.

Fig. 1 illustrates the design and possible optimal solutions of the hierarchical optimization framework for a single horizontal wellbore. The two-level structure first selects a horizontal wellbore trajectory as described in the following section and then places HF stages in the way that maximizes the NPV function for this particular well location. Once the maximum value of NPV is approximated, the next wellbore trajectory is evaluated. Though demonstrated with one well only, this framework can be easily applied to any number of horizontal wells. Later we corroborate this statement with several detailed test examples with two wells. Below, we introduce our detailed workflow for horizontal well placement and spacing optimizations of traverse HF stages.

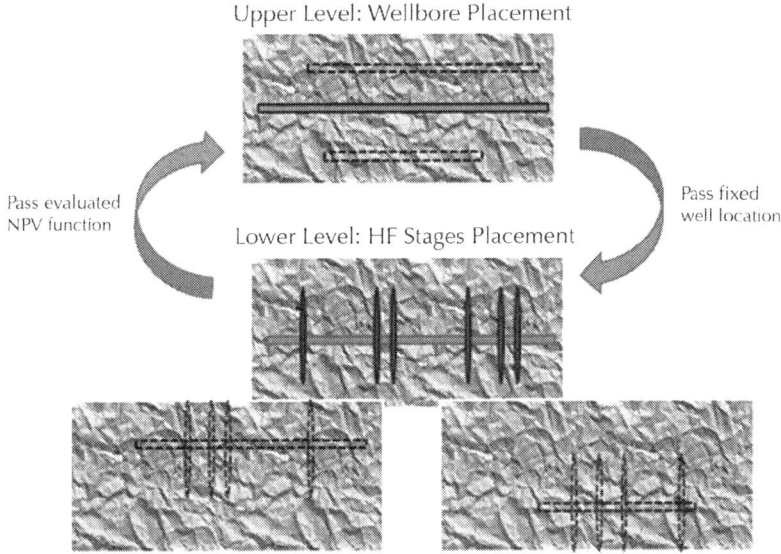

Figure 1: Conceptual model of the two-level optimization framework in fixed HF stages case.

To solve the discrete optimization problem of interest, we employ three stochastic algorithms: SPSA, GA and CMA-ES. First, all the algorithms are used to place two horizontal wellbores with fixed numbers and locations of HF stages along them. Second, with the same numerical methods we optimize locations of all HF stages. Finally, we combine these two optimization problems in one hierarchical workflow with two levels. The upper level calculates the horizontal well trajectory and the lower level places HF stages. Each algorithm is tested using both homogeneous and heterogeneous permeability maps. Results of the test runs reveal advantages and shortcomings of each combination of algorithms.

Numerical Optimization Algorithms

To finish the methodology overview, we summarize all optimization algorithms with short descriptions and schematic figures. For implementation of all numerical methods we chose the programming language MATLAB (Matlab, 2012) that was coupled with a commercially available numerical fluid flow simulator ECLIPSE (ECLIPSE, 2012).

Simultaneous Perturbation Stochastic Algorithm (SPSA)

The simultaneous perturbation stochastic approximation (SPSA) (Spall, 1992) method is particularly efficient for solving high-dimensional problems because it requires relatively small number of evaluations of the objective function in order to converge a reasonably "good" solution. SPSA uses approximation of the function's gradient rather than a direct gradient calculation which is appealing for multi-dimensional optimization problems with complex gradients.

Unlike simple finite difference method (Nocedal and Wright, 1999), SPSA works with an approximation of the gradient obtained from a stochastic perturbation of the control vector and not the gradient's actual value. By choosing a "proper" random direction k from a distribution (such as Bernoulli) in the search space of each iteration, we can find an ascent direction by computing the two-sided simultaneous perturbation using the selected random direction. Thus, the entire gradient vector is calculated with two function calls. In comparison to the simple finite difference method that requires one perturbation and forward function call for each variable, this computational efficiency is one of the biggest advantages of SPSA. SPSA requires only two function calls regardless of the size of the control vector. For the control vector with N variables and the objective function $J(u)$, SPSA approximates the gradient as follows:

$$g_i = \frac{J(u + \Delta u_i) - J(u - \Delta u_i)}{\Delta u_i}, \quad i = 1, \ldots, N \tag{3}$$

Here u_i is a perturbation vector that could be generated randomly. On average though, it nearly tracks the optimization solution because the gradient approximation is an almost unbiased estimator of the gradient, as shown in Eq. (3).

In its original form, SPSA algorithm can only operate on unbounded continuous sets and is not suitable for discrete optimization. The modified SPSA algorithm has been proposed and analyzed to the discrete well placement problem (Bangerth, 2006). In this paper, we implement and evaluate the SPSA algorithm for this discrete problem, with the goal of extending the application of the SPSA to the mixed integer problem of combined well placement and HF stages control optimization (Fig. 2).

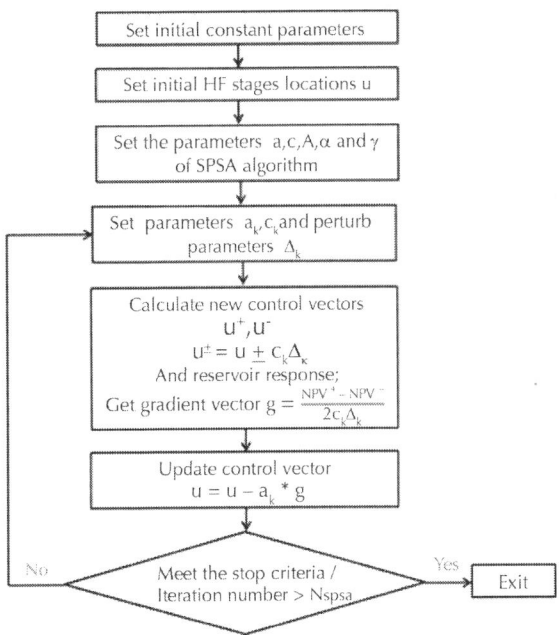

Figure 2: SPSA flowchart.

Genetic Algorithm (GA)

Gradient-based algorithms are not the only numerical tools available for optimization purposes. We can also take advantage of gradient-free methods that mimic evolution or functioning of some biological systems. The first of such algorithms is Genetic Algorithm (GA).

GA is a stochastic optimization method that uses evolutionary strategy of natural selection (Holland, 1975). GA is versatile and makes no assumptions about the shape of the objective function. It is perfectly suited for integer programming problems where control vector is represented as an array of 0's and 1's. Although the algorithm does not guarantee convergence to the global maximum in finite number of iterations, each next solution is expected to be as good as or better than one in the previous generation. Finally, as dimensions of the problem increase (the size of the control vector), slight adjustments in the GA's number of generations and individuals within generations allow preserving uniform sampling of the domain.

GA's numerous advantages explain our choice of this method for optimization of HF stages placement. This study is the first attempt to apply GA to placement of unevenly spaced HF stages. Evolutionary algorithms like GA were previously successfully applied to similar integer programming problems such as placement of horizontal wellbores (Yeten, 2003). It can be schematically summarized in the following flowchart (Fig. 3). In our implementation of GA we use maximum number of generations as the primary termination criterion and analyze performance of the algorithm by varying the maximum number of generations and observing the effect on the NPV function value.

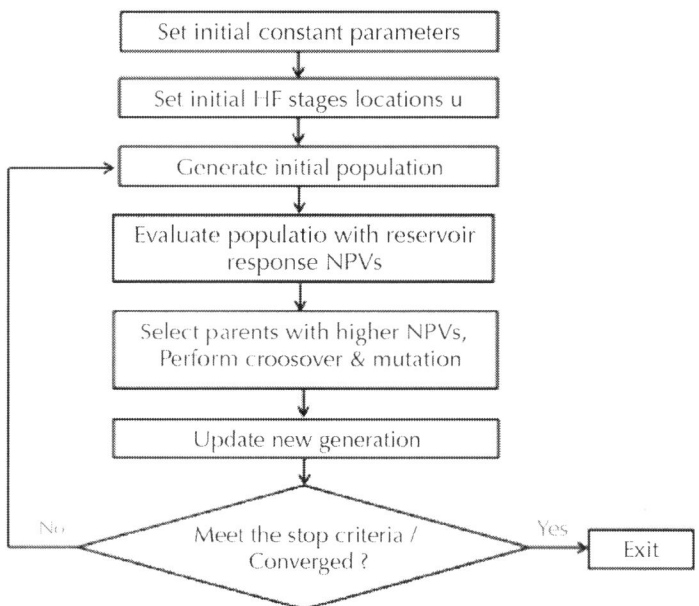

Figure 3: GA flowchart.

Covariance Matrix Adaptation–Evolutionary Strategy (CMA-ES)

Another gradient-free algorithm that we apply and test in this study is CMA-ES. Similar to GA, CMA-ES is a global optimization method that is suited for non-linear non-convex problems for which the shape of the objective function can be complex with rugged search landscape (Hansen, 2006).

Two main principles for the adaptation of parameters of the search distribution are exploited in the CMA-ES algorithm. First, there is a maximum-likelihood principle based on the idea of increasing the probability of successful candidate solutions and search steps. Second, two paths of the time evolution of the distribution mean are called search or evolution paths.

CMA-ES can be used for integer programming problems such as well placement and HF stages optimization. Therefore, the control vector for optimization by CMA-ES consists of discrete values of 1's and 0's. Fig. 4 provides the details of the CMA-ES's structure and its application to HF stages placement problem. The difference between CMA-ES and GA is that the former has covariance matrix adaptation to control the population update direction.

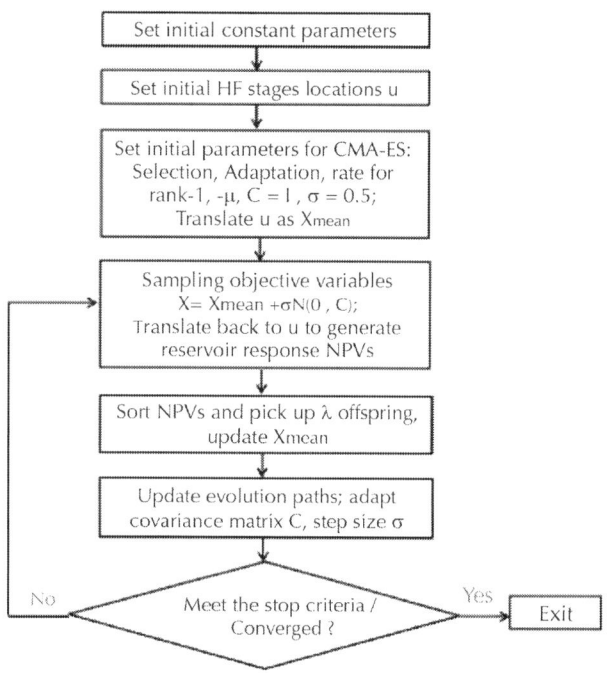

Figure 4: CMA-ES flowchart.

Flowchart of Optimization Framework in Unconventional as Reservoir

In order to start the comparative study between these three algorithms (SPSA, GA, and CMA-ES), we introduce the flowchart of our general workflow in Fig. 5. The workflow (Fig. 5) starts with setting initial conditions and defining the control vector u. For different algorithms described above, we choose different perturbation strategies to compute the gradient. SPSA chooses the Bernoulli distribution and obtains two new control vectors u^+, u^- that yield the reservoir response NPV^+, NPV^- for further gradient calculation. In GA application, multiple individual solutions are randomly generated to form an initial population. After calculation of the reservoir responses (NPVs), a proportion of the existing population is selected to breed a new generation. CMA-ES gets a group of samples around the control vector X_{mean} which also yields a number of NPVs, sorts them all and selects a certain number of control vectors with the largest NPVs to update the control vector X_{mean}. Then the process updates the control vector and repeats the iterations by each algorithm until it satisfies the termination criteria. We will show the optimization test cases and results by each algorithm in the following sections.

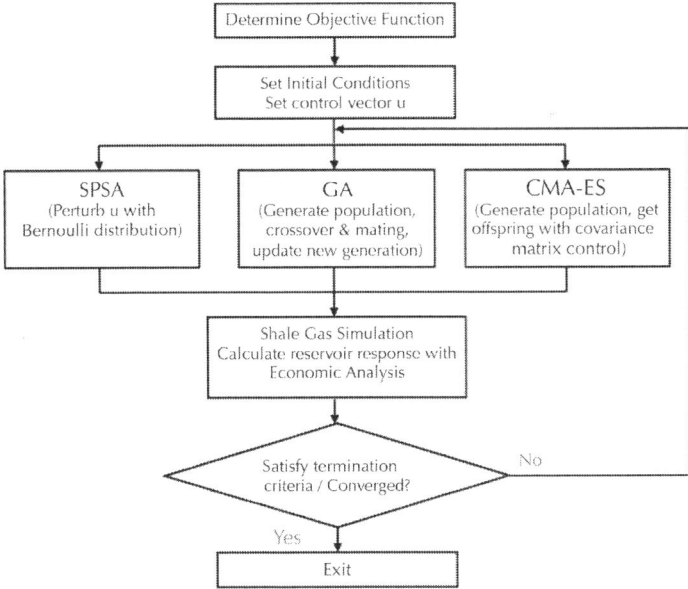

Figure 5: Flowchart of the framework for optimization applied the algorithms.

SHALE GAS MODEL

Because our objective is to find optimal locations of HF stages in shale gas reservoirs, a realistic, consistent, and practical reservoir model is required, in combination with an efficient reservoir simulator. The shale gas reservoir model in our optimization routine is simulated using a commercial reservoir simulator, namely Schlumberger compositional ECLIPSE™ 300 (E300) reservoir simulator (version 2012.2).

In the conventional reservoir area, the reservoir simulator plays an important role to do history matching and optimization works, such as inject/produce well placement and determine certain reservoir properties (Oliver et al., 2008 and Zhang et al., 2010). In the unconventional reservoir area, the model of hydraulic fracture stages and network design is also bring into effect by these simulators (Fazelipour, 2010). Shale gas reservoir model used in this study has properties from the Barnett Shale (Bruner and Smosna, 2011, Al-Ahmadi and Wattenbarger, 2011 and Mengal and Wattenbarger, 2011). In order to represent natural fracture network and capture the interaction between matrix and fracture subsystems, we apply the dual permeability model (Pruess, 1999). This shale model assumes that natural gas is stored in a local macro-porosity system (fracture porosity) and that a significant portion of the reservoir rock can be easily fractured (Fig. 6).

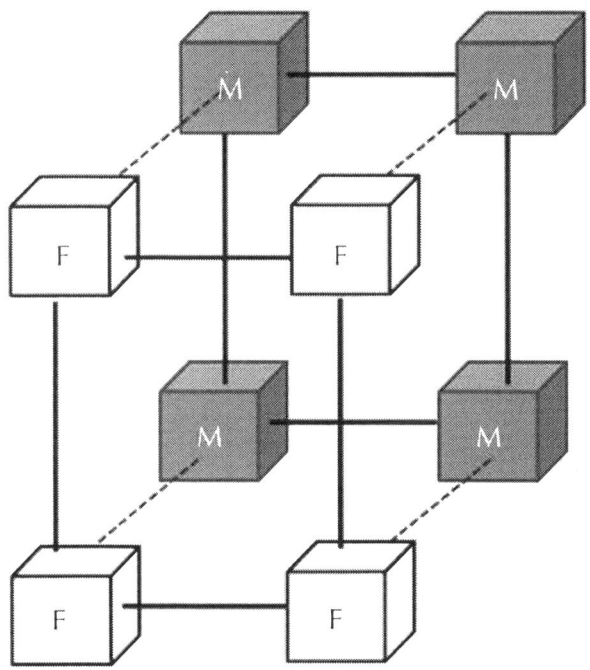

Figure 6: Flow connections in the dual permeability model (adapted from Pruess, 1999). *M* corresponds to matrix permeability and *F* to fracture permeability.

The dual permeability flow system describes the flows from matrix to matrix cells, from fracture cell to fracture cell, and from matrix cell to its corresponding fracture cell and vice versa. The use of the dual permeability model with flow from matrix to matrix cell is common and has been successfully applied to coal bed methane reservoirs (Pruess, 1999). In the dual permeability model, the matrix cell of a matrix-fracture coupled gridblock is treated as a source term, which has adsorbed gas in the organic materials. The source, upon an applied pressure drawdown, expels gas as free gas into the matrix porosity and subsequently the fracture network cell, linked within the same matrix-fracture coupled gridblock. The fracture network cell acts as a sink term in this process. The Langmuir isotherm parameter is widely used to describe gas adsorption, as shown in Eq. 4

$$V(P) = \frac{V_L \cdot P}{P_L + P},\qquad(4)$$

where $V(P)\left[\frac{Mscf}{ton}\right]$ is the adsorbed gas content at pressure $P[psi]$, V_L is the Langmuir volume parameter [Mscf/ton] which gives the storage capacity of adsorbed gas content at infinite pressure, and P_L the Langmuir pressure parameter [psi]. The calculation also needs the bulk density of shale ($_{bulk}$) to convert typical gas content from [Mscf/ft3] to [Mscf/ton]. The adsorption data from Mengal and Wattenbarger (2011), as depicted in Fig. 7, is used as the Langmuir isotherm of the Barnett Shale reservoir problem in this this paper.

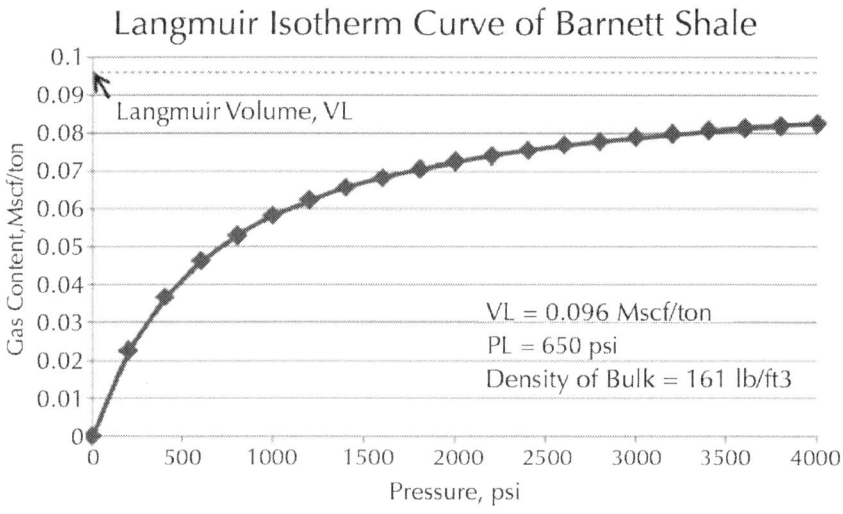

Figure 7: Langmuir isotherm curve and adsorption data of Barnett Shale.

We have used the reservoir simulator ECLIPSE E300 module to build up on the reservoir model using the Barnett Shale (Al-Ahmadi, 2011) properties in Table 2. The matrix permeability and the nature fracture permeability (marked as *) are initially unknown and were determined after history matching with the field data.

Table 2: Reservoir properties and hydraulic fracture parameters used in history matching

Parameters	Values	Unit
Model dimensions	3000 × 1510 × 300	ft
Initial reservoir pressure	2950	psi
Reservoir temperature	150	F
Bulk density	161	lbs/ft^3
Bottom hole pressure	500	psi
Horizontal well length	2968	ft
Production period	5	years
Matrix permeability	0.00015*	md
Matrix porosity	0.06	100%
Natural fracture efficient permeability	0.0001	md
Natural fracture porosity	0.00005*	100%
Hydraulic fracture conductivity	1	md-ft
Hydraulic fracture spacing	100	ft
Hydraulic fracture height	300	ft
Hydraulic fracture half-length	105	ft
SRV permeability, Zone 1	0.05	md
SRV permeability, Zone 2	0.0005	md
Number of hydraulic fractures	28	Stages

*Indicates history matched parameters based on the real field data from Barnett Shale.

In this model, the reservoir is assumed to be homogeneous and the multistage HFs is evenly spaced along the horizontal well with a single perforated interval for each stage (Fig. 8). Fig. 9 provides the results from the history matched model against field production data from well 314 of the Barnett Shale field data. FromFig. 9, one can observe that there is a good agreement between the numerical model used in this paper and the real field data, which also provide a reasonable simulation model for the production prediction and optimization approaches to be performed in the next sections.

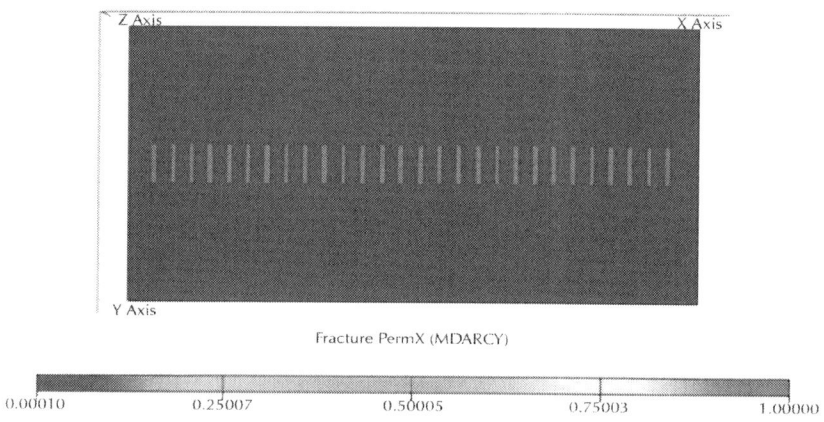

Figure 8: Model of multistage hydraulic fractures distribution along horizontal well.

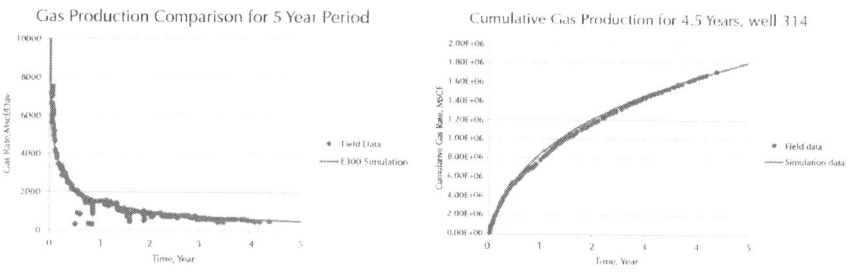

Figure 9: Gas production history matching of well 314 from Barnett Shale.

Because the dual permeability model allows accounting for contribution from the natural fracture subsystem, we devise test cases with both homogeneous and complex heterogeneous permeability maps.Fig. 10 presents one test example which models a shale reservoir with complex network of natural fractures. To finish the overview of the shale gas model, we specify gridding and representation of HF stages. The shale gas model under consideration has a coarse grid with cell dimensions of 20 × 20 × 200 ft. To represent traverse HF's more accurately within the model, we enable local grid refinement feature (LGR) in particular coarse gridblocks. The LGR's contain nine subdivisions with different width ratios. The smallest subdivision at the center has the width of only 0.4 ft and HF permeability. A zone representing stimulated reservoir volume (SRV) around each HF with enhanced permeability is also incorporated into the model (Fig. 11). LGR's and SRV's change automatically as HF's switch their locations during the optimization process.

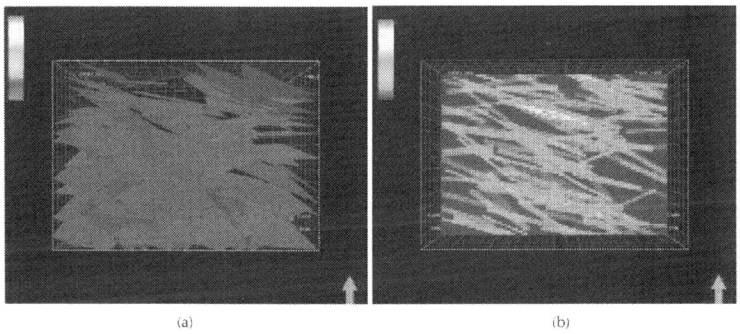

Figure 10: (a) Fracture network generated by Petrel; (b) permeability map after upscaling.

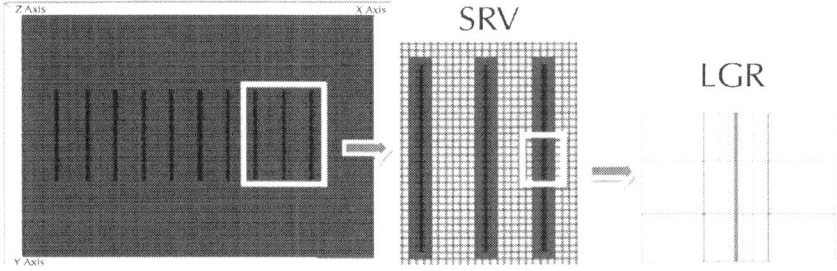

Figure 11: LGR and SRV features used in the model.

NUMERICAL EXPERIMENTS RESULTS

In this section, we present results of several numerical experiments in order to evaluate the performance of all these three algorithms. We define several suites of simulation runs to test optimization algorithms and observe performance on geo-models with homogeneous and heterogeneous permeability maps. In addition to this, we vary initial conditions from case to case and compare the results of three optimization tasks: well placement, HF stages placement, and hierarchical framework for integrated well and HF stages placement.

Well Placement Optimization

In wellbore placement optimization we assume that two horizontal wells are placed in the reservoir in the direction from west to east. To apply these algorithms to wellbore placement problem, we define the control vector u in terms of x-, y-, and z-coordinates of the toe and the heel. By fixing the wellbore length at 2000 ft and keeping two horizontal wells parallel, the optimization problem can be simplified to find the optimal distance between the two wellbores. Mathematically, this problem is formulated with the following constraints:

$$\begin{cases} \max J(u) \\ s.t. Y_{max} \geq Y(u) \\ Y_{min} \leq Distance(Y(u)) \end{cases} \quad (5)$$

These constraints are necessary to guarantee the existence of the optimal solution within the feasible domain. The first constraint requires each well to stay inside the reservoir grid and keep some distance to its boundaries. The second one dictates the minimal distance between the wells which could build up an effective fracture network for long-term production period.

In this optimization task each wellbore has ten HF stages with fixed locations (Fig. 12). For the case with homogeneous permeability map, each algorithm is tested five times with different initial wellbore arrangements. These arrangements are generated with different random seeds from the uniform distribution. The results demonstrate that we find the optimal wellbore locations and maximum value of NPV within several iterations (Fig. 14). Rapid convergence of both algorithms can be explained by the size of our problem: the control vector u in this case has only two dimensions. Fig. 13 demonstrates the importance of the optimal spacing between the wellbores for improved reservoir production and uniform pressure decline. Fig. 14(d) gives the juxtaposition between improved NPVs obtained by different algorithms; each NPV curve is the average value calculated from the corresponding algorithm curves shown in Fig. 14(a–c). Same format as the other following cases. Comparison between these optimization results yields the following observations: all the algorithms converge fast to the same NPV value and obtain the same wellbore trajectories corresponding to this NPV (Fig. 12 and Fig. 14). In the homogeneous cases, NPVs are improved around 5%. The improved NPV and comparisons between tests in heterogeneous cases are listed in Fig. 15 and Fig. 16 and Table 3.

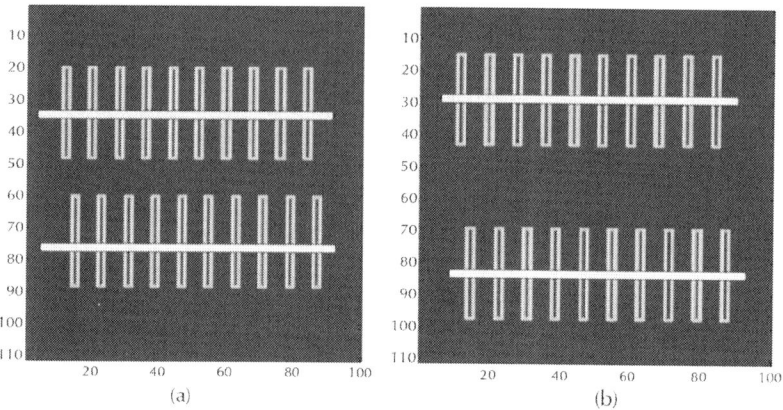

Figure 12: Homogeneous permeability map: (a) one of initial wellbore placement and (b) the optimized result of wellbore locations.

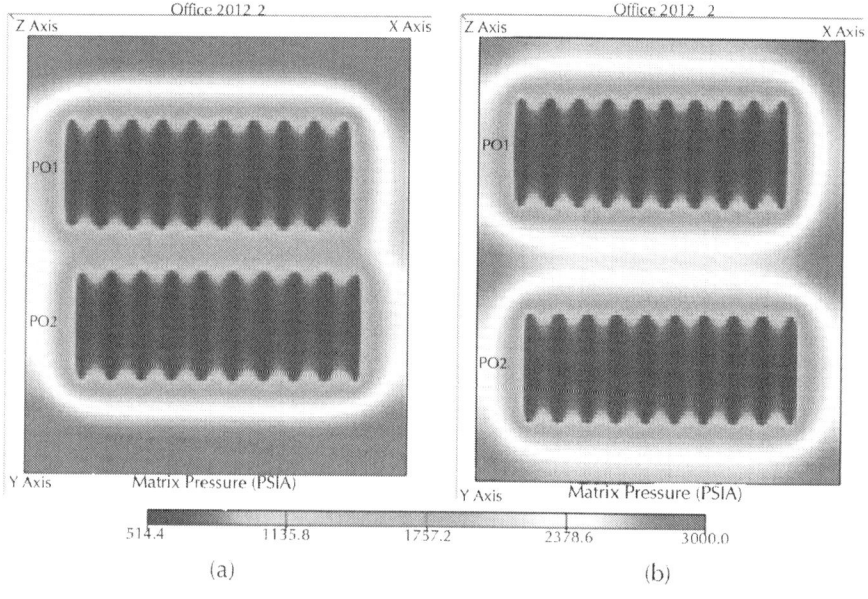

Figure 13: Pressure distribution after 20 years production for (a) the initial case and (b) the optimized results.

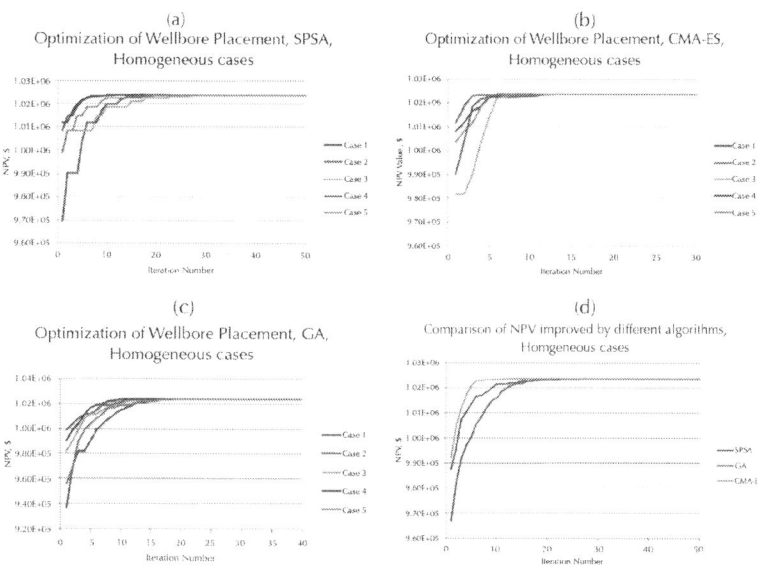

Figure 14: Wellbore placement optimization for homogeneous permeability map.

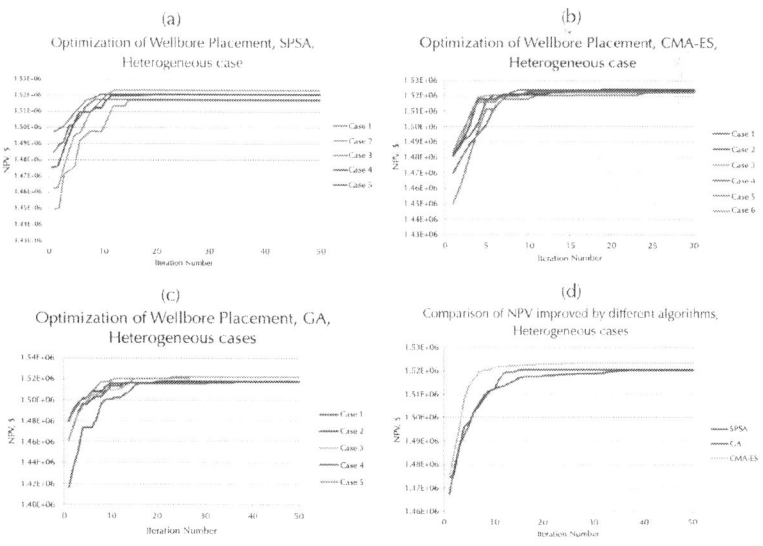

Figure 15: Wellbore placement optimization for heterogeneous permeability map.

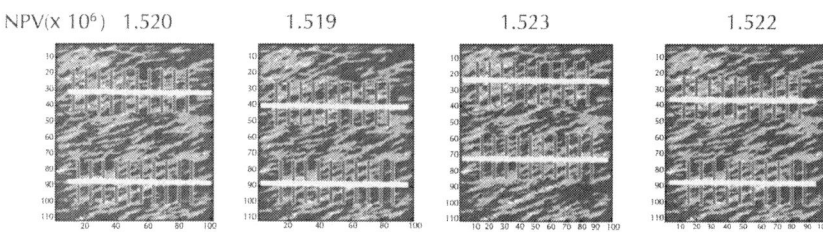

Figure 16: The optimization distributions of wellbore placement for four different initial using the known geologic model in heterogeneous case.

Table 3: NPVs of four wellbore placement cases in heterogeneous case in Fig. 16

Cases	(a)	(b)	(c)	(d)
Improved NPV ($, 10^5)	1.520	1.519	1.523	1.522
% of improved NPV	4.11%	4.08%	4.51%	5.23%
Optimization result	(31, 89)	(38, 90)	(36, 89)	(26, 72)

In the case of the heterogeneous permeability map, we observe around 10% increases in NVP by each algorithm (Fig. 15(a–c)). All algorithms converge to the solution within several iterations, but CMA-ES yields slightly higher optimal value and obtains it with smaller uncertainty than the other two algorithms (Fig. 15(d)). In all the tests, we use the same heterogeneous permeability map with different initial wellbore arrangements, and some of the optimization results are given in Fig. 16. Although the final optimal NPVs are close, most of them are the local optimization solutions. SPSA's advantage becomes apparent in this test suite: SPSA requires the smallest number of the simulator calls and total computation time. CMA-ES performs slightly better than GA: CMA-ES gives higher improved NPV and requires smaller number of iterations to converge to the optimal results because it has the covariance matrix to control the new generation direction and evolution paths.

HF Stages Placement Optimization

We first test a case with a single horizontal well placed in the reservoir's middle zone in the direction from west to east. In HF stages placement optimization we try to establish the optimal locations of the stages along the wellbore. This goal is achieved by putting the following constraints on the variables of the control vector u:

$$\begin{cases} \max J(u) \\ s.t. l_{toe} \leq u_1 \\ u_n \leq l_{heel} \\ l_{min} \leq u_i, \ i = 1, 2, \ldots, n \end{cases} \quad (6)$$

Eq. (6) describes the feasible domain of the problem by, first, confining all HF stages to the gridblocks along the well trajectory (from l_{toe} to l_{heel}) and, second, keeping some minimal distance l_{min} between the stages. l_{min} is the minimum HF stage spacing that is around 400 ft based on economic analysis from available shale field data. In this optimization problem the control vector u increases in dimensions, because now it records the intervals with corresponding grid numbers between all HF stages.

Initially the wellbore has ten HF stages with unknown (or evenly spaced) locations, which are shown in left figure of Fig. 17. Because of the stochastic nature of the numerical algorithms, we run five times with the same initial conditions and compare the results as follows. Fig. 18 illustrates that, although the optimization problem is multi-dimensional, each algorithm converges within a small number of iterations. The improved NPVs obtained by different algorithms are very close. This observation demonstrates the ability of the three algorithms to handle the large number of variables in the control vector. Similarly to the wellbore placement problem, CMA-ES gives the highest value of the NPV function while

SPSA is the most efficient in terms of simulator calls and overall computation time.

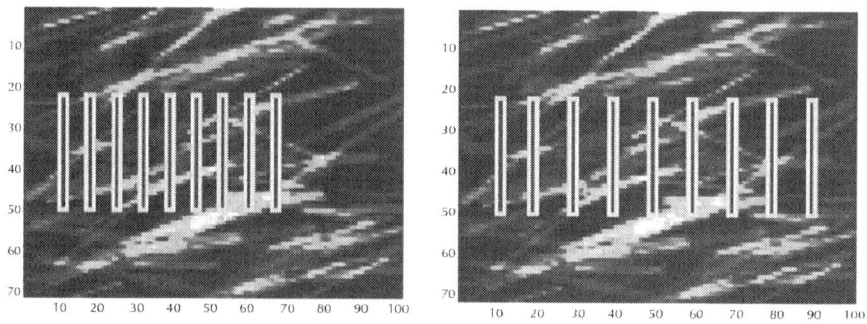

Figure 17: Initial HF stages placement on single well (left) and the optimized result (right).

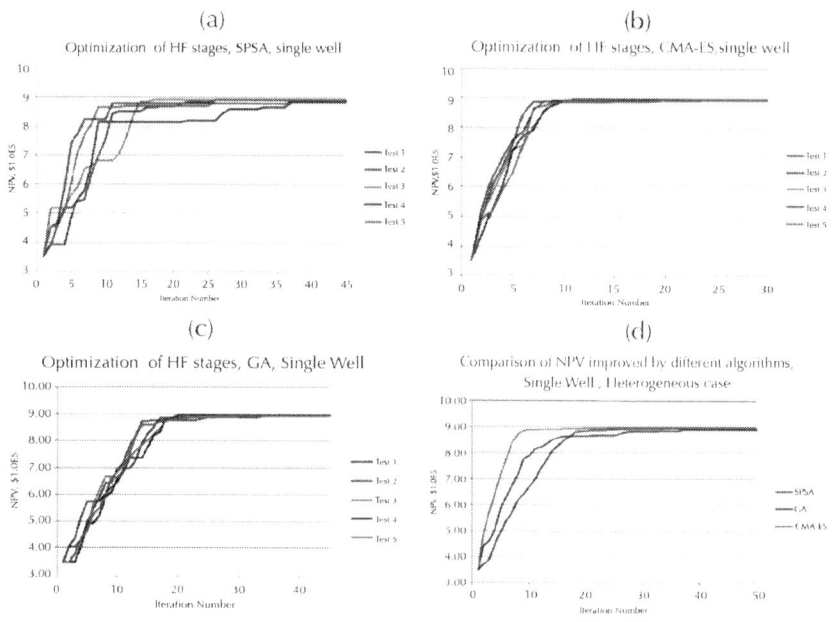

Figure 18: HF stage placement optimization for a single well.

In this section, we consider test cases with two horizontal wells placed in the reservoir in the direction from west to east. Initially each well has ten HF stages with unknown (or evenly spacing) locations. The control vector for this optimization problem has two times more dimensions than that for the single well case. Here we move locations of HF stages along the two wellbores simultaneously.

Fig. 19 gives the initial uniform distribution of HF stages along two fixed wellbores and the optimal distribution of the stages after application of our optimization approach. After optimization, NPV improves by about 25% in comparison to conventional evenly spaced technique. Fig. 20 shows that all the algorithms have the capability of solving problems with the large number of dimensions. CMA-ES and GA obtain better solutions with higher total computational times because both algorithms use a large number of offspring ( ). The large offspring population, however, produces the near optimal solutions with less uncertainty in comparison to SPSA. The heterogeneous permeability map case in Fig. 21 and Fig. 22 also supports the same observations. Table 4 lists the percentages of NPV improvement and, thus, demonstrates obvious advantage of our optimization process from the economical point of view. Comparison between these algorithms' results reveals that CMA-ES obtains the optimal solution with less uncertainty and its NPV values are higher than those from the other algorithms. This illustrates that CMA-ES and GA are global optimization methods, while SPSA tends to get trapped in local optima.

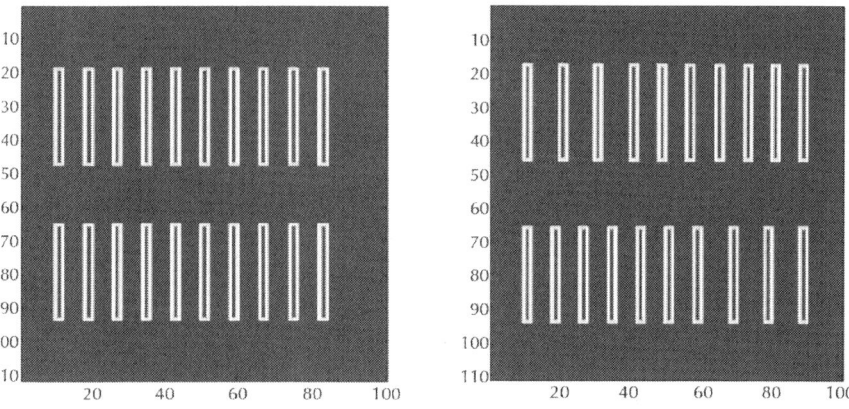

Figure 19: Homogeneous case, initial HF stages placement on two wells (left) and one optimized result (right).

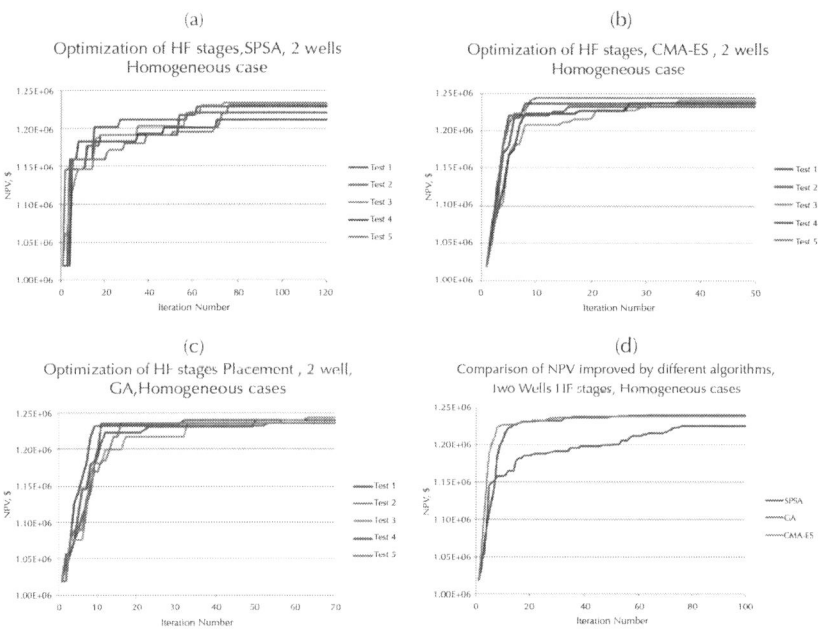

Figure 20: HF stages placement optimization for two wells in homogeneous permeability map.

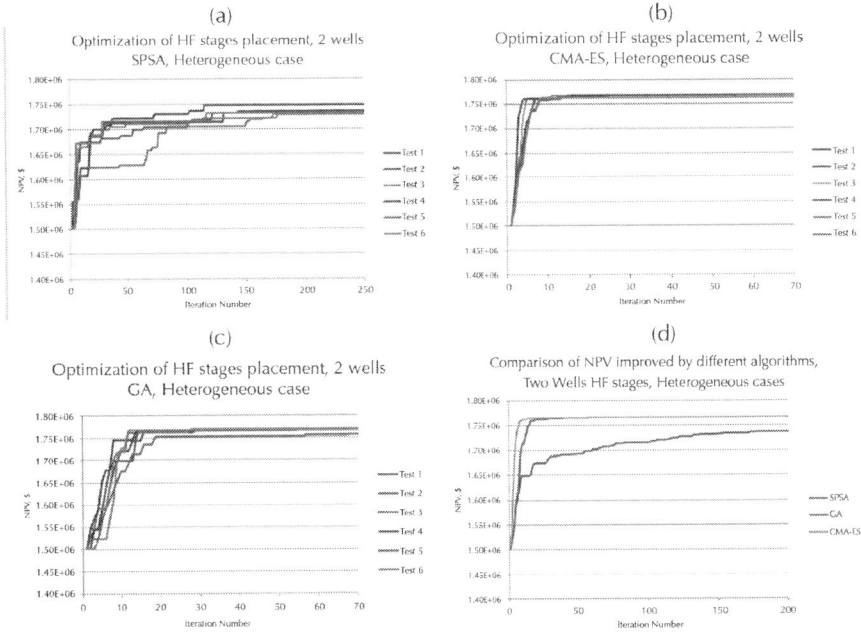

Figure 21: HF stages placement optimization for 2 wells in heterogeneous permeability map.

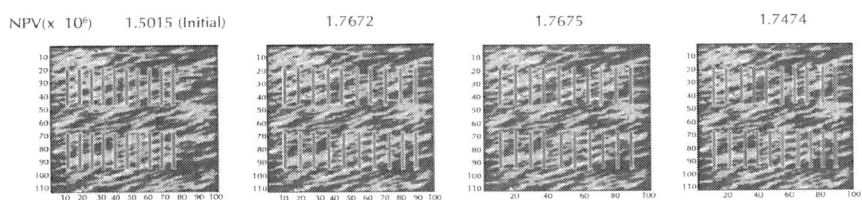

Figure 22: Initial case (left) and three optimization distributions of HF stages placement using the known geologic model in heterogeneous case.

10. Holditch, S.A., 2007. Hydraulic fracturing: overview, trends, issues. Fractur. Stimul. 2007, 26–28.
11. Holt, S., 2011. Numerical Optimization of Hydraulic Fracture Stage Placement in a Gas Shale Reservoir. M.S. Thesis. TU Deflt University, Netherlands. Holland, J.H., 1975. Adaptation in Natural and Artificial Systems: An Introductory Analysis with Applications to Biology, Control, and Artificial Intelligence. The University of Michigan Press, Ann Arbor, Michigan, p. 211.
12. Jenkins, C.D., Boyer, C.M., 2008. Coaled and shale gas reservoirs. J. Petrol. Technol. 60 (2), 92–99, SPE 103514.
13. MATLAB version R2012b. 2012. The MathWorks Inc., Natick, Massachusetts. Mengal, S.A., Wattenbarger R.A., 2011. Accounting For Adsorbed Gas in Shale Gas Reservoirs. SPE Middle East Oil and Gas Show and Conference, Manama, Bahrain, SPE 141085, p. 15.
14. Nocedal, J., Wright, S.J., 1999. Numerical Optimization. Springer Series in Operations Research, New York, USA, p. 636.
15. Oliver, D.S., Reynolds, A.C., Liu, N., 2008. Inverse Theory for Petroleum Reservoir Characterization and History Matching. Cambridge University Press, New York, USA, p. 394.
16. Pruess, K., Oldenburg, C., Moridis, G., 1999. TOUGH2 User's Guide, Version 2.0., Earth Sciences Division, Lawrence Berkeley National Laboratory. Berkeley, CA, USA. LBNL-43134.
17. Schweitzer, R., Bilgesu, H.I., 2009. The Role of Economics on Well and Fracture Design Completions of Marcellus Shale Wells. SPE 125975.
18. Spall, J.C., 1992. Multivariate stochastic approximation using a simultaneous perturbation gradient approximation. IEEE Trans. Autom. Control. 37, 332–341.
19. Yeten, B., 2003. Optimum Deployment of Nonconventional Wells. Ph.D. dissertation, Stanford University, CA. Available

at: <https://pangea.stanford.edu/ERE/pdf/ pereports/PhD/Yeten03.pdf>.
20. Yu, W., Sepehrnoori, K., 2013. Optimization of Multiple Hydraulically Fractured Horizontal Wells in Unconventional Gas Reservoirs, SPE Production and Operations Symposium, Oklahoma City, Oklahoma, SPE 164509.
21. Zhang, K., Li, G., Reynolds, A.C., Yao, J., Zhang, L., 2010. Optimal well placement using an adjoint gradient. J. Petrol. Sci. Eng. 73, 220–226.

Chapter 4

A Reinterpretation of Historic Aquifer Tests of Two Hydraulically Fractured Wells by Application of Inverse Analysis, Derivative Analysis, and Diagnostic Plots

Patrick A. Hammond[1] and Malcolm S. Field[2]

[1]Source Protection & Appropriation Division, Maryland Department of the Environment, Baltimore, USA (Retired)

[2]National Center for Environmental Assessment, US Environmental Protection Agency, Washington DC, USA

ABSTRACT

Aquifer test methods have greatly improved in recent years with the advent of inverse analysis, derivative analysis, and diagnostic plots. Updated analyses of past aquifer tests allow for improved interpretations of the data to enhance the knowledge and the predictive capabilities of the flow system. This work thoroughly reanalyzes a series of pre- and post-hydraulic fracturing, single-well aquifer tests conducted in two crystalline rock wells in New Hampshire as part of an early 1970's study. Previous analyses of the data had relied on older manual type-curve methods for predicting the possible effects of hydraulic fracturing. This work applies inverse analysis, derivative analysis, and diagnostic plots to reanalyze the 1970's aquifer test data. Our results demonstrate that the aquifer tests were affected by changes in flow regimes, dewatering of the aquifer and discrete fractures, and changes due to well development. Increases in transmissivities are related to well development prior to hydraulic fracturing, propagation of a single, vertical fracture hydraulically connecting the two wells after stimulation and expansion of troughs of depression. After hydraulic fracturing, the estimated total yield of the individual wells increased by 2.5 times due to the hydraulic fracturing. However, the wells may be receiving water from the same source, and well interference may affect any significant increase in their combined yield. Our analyses demonstrate the value in applying inverse analysis, derivative analysis, and diagnostic plots over the conventional method of manual type-curve analysis. In addition, our improvement in the aquifer test interpretation of the 1970's test data has implications for more reliable estimates of sustained well yields.

INTRODUCTION

To a large extent, aquifer-test analyses still rely on the type-curve matching methodology as originally introduced by C. V. Theis [1], in which a theoretical curve is manually overlain onto a measured time-drawdown curve, and the Cooper-Jacob straight-line modification of the Theis solution [2] for obtaining aquifer parameters (see for example, [3] -[5]). Often overlooked is the fact that most type curves are just variations on the original Theis type curve that are designed to account for specific aquifer characteristics that are different from those for which the Theis type curve was originally intended. With the advent of the digital computer and enhanced analytical capabilities, aquifer-test analyses show considerable improvement over the common practice of manually aligning a type curve over the drawdown data.

Application of such techniques as derivative analysis, diagnostic plots, and inverse analysis represents substantial enhancements to the typical type-curve matching methodology. Inverse analysis, the process of numerically fitting a theoretical curve to a set of measured data, has been in existence for several decades (see Chapter 6 in [6], and references therein). Derivative analysis of pumping test data relates the rate of drawdown change as a function of the natural logarithm of time. It permits assessment of changes in drawdown response not easily discerned from the drawdown data and to identify flow regimes encountered during a pumping test [7]. Diagnostic flow plots simultaneously graph drawdown against various functions of time [8]. They have been applied to identify and characterize flow regimes [8], facilitate the selection of appropriate analytical models for drawdown data, and estimate appropriate hydraulic properties of the aquifers. Derivative analysis and diagnostic plots have also been applied to determine flow dimensions in heterogeneous fractured media [9].

This paper applies a combination of inverse analysis, derivative analysis, and diagnostic plots to enhance well-test interpretations and assess changes in aquifer hydraulic behavior and productivity after hydraulic fracturing of two wells [10]. The data set has been previously analyzed using conventional manual type curve matching as part of related studies on predicting the effect of hydraulic fracturing [3] [10] . Reinterpretation of the data using more advanced techniques can provide insight about the flow regions, appropriate applicable models, and potential changes in hydraulic properties before and after hydraulic fracturing. We contend that the more advanced techniques used in this paper yield more accurate estimates of hydraulic properties than conventional curve-fitting methods, and enhances the ability to simulate the system for better aquifer management. Reevaluating the results of Stewart [10] for a more reliable assessment is especially significant because the original reported improvements in yield after hydraulic fracturing have been reprinted in later references (e.g., [11] , p. 533).

REVIEW OF ANALYTICAL METHODS

The model often used to analyze pumping test data is the two-dimensional solution developed by Theis [1], now commonly called an infinite acting radial flow (IARF) model, which assumes confined conditions in a uniform, homogeneous, isotropic aquifer, where flow is radial and there is no vertical flow component. Boulton [12] - [14], Neuman [15] - [17], Streltsova [18], and Boulton and Streltsova [19] - [22] developed analytical models that defined delayed yield and pseudo-equilibrium responses noted during tests. In those cases, the type curves are S-shaped, and are very similar to those derived for dual porosity aquifers, such as the one developed by Moench [23] which accounts for flow in a system consisting of low permeability matrix blocks and high permeability non-directional fractures. Single porosity models describe linear/pseudo-radial flow controlled by prominent directional fractures [24].

Other commonly applied models are those describing generalized radial flow [25] for spatial and dimensional flow in single or double porosity fractured aquifers and leaky confined aquifers [26] [27] overlain or underlain by constant head or no flow boundaries. These studies are not the only ones available for analyzing aquifer tests in fractured rock areas (see Chapter 10 in [28] and references therein), but a common characteristic these models have is that the type curves of most of these models merge with the Theis curve at late time. The Cooper-Jacob straight-line method [2] can often be used on a semi-log plot to evaluate time-drawdown data if that portion of a curve that represents an IARF period can be identified.

Derivative Analysis

The primary tool for the derivative analysis method is a simultaneous plot of drawdown and the logarithmic derivative of drawdown as a function of time. It is now considered the best method for identifying an appropriate conceptual model to use when analyzing aquifer test data [29]. Bourdet et al. [30] developed an algorithm for the petroleum industry that calculates the first derivative of the pressure change with respect to the natural logarithm in the change of time according to

$$\left(\frac{dp}{dx}\right)_i = \frac{\Delta p_1 \Delta x_2 / \Delta x_1 + \Delta p_2 \Delta x_1 / \Delta x_2}{\Delta x_1 \Delta x_2}, \tag{1}$$

where p is pressure, subscript 1 = point(s) before the point of interest, i; subscript 2 = point(s) after the point of interest, i; and x is the natural logarithm of the time function, t*. Because drawdown s during an aquifer test is related to pressure, Equation (1) may be applied to piezometric head $\phi(t)$ or drawdown with time s(t) [31]

$$\phi t = \int_{p_0}^{p} \frac{dp}{\rho g} + z, \tag{2}$$

$$s(t) = \phi t - \phi(t_0),\tag{3}$$

to estimate the logarithmic derivative of the drawdown (ds / dx = tds / dt) for a reference elevation z, and an elapsed time t since the beginning of the pumping test (t_0) is a suitable time. Fitting the derivative of a conventional type curve directly to the first divided differences of the observed drawdown data may be accomplished using [31]

$$\frac{\Delta s_i}{\Delta t_i} = \frac{s(t_{i+1}) - s(t_i)}{t_{i+1} - t_i}\tag{4}$$

or as described by Renard et al. [8]

$$\frac{\Delta s_i}{\Delta \ln t_i} = \frac{(t_i + t_{i+1})}{2} \frac{\Delta s_i}{\Delta t_i}\tag{5}$$

Derivative analysis of drawdown is applied to single-well tests to assess the influence of well-bore storage, type of aquifer, presence of boundaries, and flow regimes in the data [8]. Different flow mechanisms or behaviors can occur throughout the pumping period as the expanding drawdown encounters boundaries and heterogeneities. This requires that different phases of an aquifer test be analyzed separately. For instance, if an IARF segment can be identified on a semi-log plot, which occurs when the derivative stabilizes at a constant level, then the Cooper-Jacob straight-line solution can be applied. If IARF is not present, then the results of a derivative analysis can be used to determine which other conceptual models should be used to evaluate the data from a test.

Renard et al. [8] provides a synthesis of the behaviors of typical drawdown and log derivative plots in response to constant pumping rates. Different flow mechanisms or behavior can be combined, giving rise to several conceptual models to evaluate the data. The models represent typical behavior of representative aquifer conditions, including: infinite two-dimensional confined aquifer (IARF); double porosity or unconfined aquifer; infinite linear no-flow boundary; infinite linear constant-head boundary; leaky aquifer; well-bore storage and skin effects; infinite conductivity vertical fracture; general radial flow with flow dimensions smaller or larger than 2; and the combined effect of well storage and infinite linear constant head-boundary [8].

Although the derivative analysis method is considered the best method for identifying an appropriate model for aquifer test analysis, it requires a large number of calculations that are best handled by computer generated algorithms. The AQTESOLV program (version 4) uses several diagnostic or specialized flow plots to aid in the identification of different flow regimes. These include: radial flow (s vs. t), linear flow (s vs. $t^{1/2}$), bilinear flow (s vs. $t^{1/4}$), and spherical flow (s vs. $t^{-1/2}$). Radial flow models apply to homogeneous systems in which IARF is radially symmetric around the well [9]. Drawdown data in these systems are typically described by the Theis type curve and shows constant positive drawdown and derivatives at late times [8]. Linear flow is associated with single, infinite, or high conductivity fractures and channel strip aquifers. They tend to show a 1:2 slope in the drawdown and derivative plots at different times during the pumping test [7]. Single fractures show a 1:2 slope at early times followed by IARF, whereas channel strip aquifers reflect the 1:2 slope at late times. Bilinear flow is associated with single, finite, or low conductivity vertical fractures represented by 1:4 slopes at early times. Spherical flow is associated with partially penetrating wells and tests conducted on packed intervals. Other flow mechanisms that can be detected using diagnostic and derivative radial flow plots include: well-bore storage effects (initial unit slope, followed by a drop-out, forming a peak on the plot), closed aquifer (late-time unit slope), water table and dual porosity aquifers (dip in the drawdown at mid-time), leaky aquifers (initial positive drawdown, followed by an approach to zero drawdown at late-time), recharge boundaries (constant zero drawdown), and impermeable barriers (two levels with constant positive drawdowns). Duffield [7] discusses in detail methods for the performance and interpretation of derivative analyses when using the AQTESOLV program.

Inverse Analysis

The most useful tool for determining aquifer properties and estimating reliable yields of wells is the aquifer pumping test. The common graphical method of fitting type curves to data on a log-log plot to derive aquifer constants provides simple solutions to inverse problems, but can be subjective in nature and is prone to errors in individual judgment. The primary reason for these errors is that type curves representing different flow mechanisms often have such similar shapes that each can provide relatively good visual fits to the same set of data. Computer assisted inverse analysis techniques, the process by which a theoretical curve is numerically fitted to a data set is generally regarded as less subject to individual bias. However, choice of incorrect models, for example, may still result in acceptable model fits to a data set even though the results are incorrect [32]. The possibility of choosing an incorrect model based on an inadequate knowledge of site hydrogeology led Johns et al. [33] to state that the graphical method for aquifer test analysis represents a compliment to inverse analysis of aquifer tests.

Estimating aquifer parameters and extrapolating drawdown data by inverse analysis ([6], p. 229) can remove much of the subjective nature involved with graphical methods. It is, however, essential that the inverse problem be well-posed so that the solution is unique and stable [34] ([35], p. 12). Typically, the inverse problem can be solved using such methods as the Levenburg-Marquart method (e.g., [36] , pp. 24-29), but if multiple peaks exist in the data causing a global minimization problem, then other methods such as genetic algorithms (e.g., [37]) may be needed.

Inverse analysis uses a nonlinear least squares procedure that seeks to minimize the sum of the squared residuals (RSS) between the measured data and the theoretical type curve drawdowns for the analytical model under consideration. The first known use of inverse analysis for aquifer test analysis was provided in Saleem [38], although the method had previously been applied to petroleum reservoir analysis [39] [40]. A significant number of other studies on the subject were developed over the following 20 years, and many are included as references in the Sayed [41] and Johns et al. [33] investigations. Most of the published work consists of programs designed to match test data to one or a few analytical models. The commercially available AQTESOLV program [7], includes 35 different analytical models, most of which can be applied to fractured rock aquifers. It is used to evaluate the aquifer test data in the present study. Manual or visual curve fitting techniques may be used in the AQTESOLV program to provide preliminary estimates of hydraulic properties, which may improve the automated process of matching simulated to observed values.

EXAMPLE ANALYSIS

This paper applies inverse analysis, derivative analysis, and diagnostic plots to analyze a series of step and single-well aquifer tests conducted by Stewart [10] before and after hydraulic fracturing of two wells in New Hampshire. Although single-well aquifer tests often result in appreciable errors [42] in their ability to represent aquifer properties they may still provide reasonably acceptable results. Other similar studies have been conducted in Australia, Zimbabwe, and Canada to assess the effect of hydraulic fracturing on well yield [43] - [45], but they do not present a complete set of geophysical, aquifer test, or hydraulic stimulation data. Stewart [10] presents complete sets of these data for tests conducted in the same wells before and after hydraulic fracturing.

Two wells (A and B) were used for the aquifer tests presented in the Stewart [10] study. The wells were located 207 m apart (Figure

1) at the University of New Hampshire horticultural farm, Durham, New Hampshire.

Well A was drilled 146 m into granitic bedrock throughout its full depth. Well B was drilled 91 m into schistose rock, with 17 m of overlying surficial material consisting mostly of glacio-marine silts and fine sands. Each well was hydraulically fractured in an effort to increase productivity.

Hydraulic fracturing consists of injecting a highly pressurized fluid into an isolated section of a borehole to induce and propagate fractures into the rock matrix, in order to improve the productivity of a well. It was first used by the petroleum industry in the late 1940's for the stimulation of oil and gas wells [46], and later developed for the drinking water industry in the 1960's [47].

Hydraulic fracturing methods for water wells, however, require relatively lower pressures than for oil/gas wells to open existing fractures that may intersect wellbores.

Water well drillers typically report that production test rates may increase by an order of magnitude after a well has been stimulated by hydraulic fracturing. However, the reported increases are typically low volumes of water changing from substantially less to slightly greater than ~4.0 $L \cdot min^{-1}$. To the authors' knowledge, there are no known published studies that address the long-term yields of such wells. Although there are many published papers in the petroleum literature on hydraulic fracturing in oil and gas fields, few comprehensive studies had been conducted to determine the effectiveness of hydraulic fracturing on the production of water wells that might be used for public or commercial/industrial water supplies.

The single-well tests described in Stewart [10] consisted of pumping each well separately at a constant rate while monitoring water levels. Different rates and pumping periods were used for different tests. Seven constant rate pumping tests were reported for each well (Tests A2, A3, A5, A14, A16, A741, A742, B3, B4, B25, B27, B28, B743 and B744). Tests A2, A3, A5, and B3 and B4 were conducted prior to hydraulic fracturing, whereas Tests A14,

A16, A741, A742, B25, B27, B28, B743 and B744 were pumped afterwards. The hydraulic fracturing was focused on the bottom 27 m of well a (depth 116 - 146 m) and on the bottom 15 m of Well B (depth 76 - 91 m). Two additional aquifer tests conducted in Well B failed due to pump seizures, and were not reported. Four series of step-drawdown tests were also described in Stewart [10], but no data is presented.

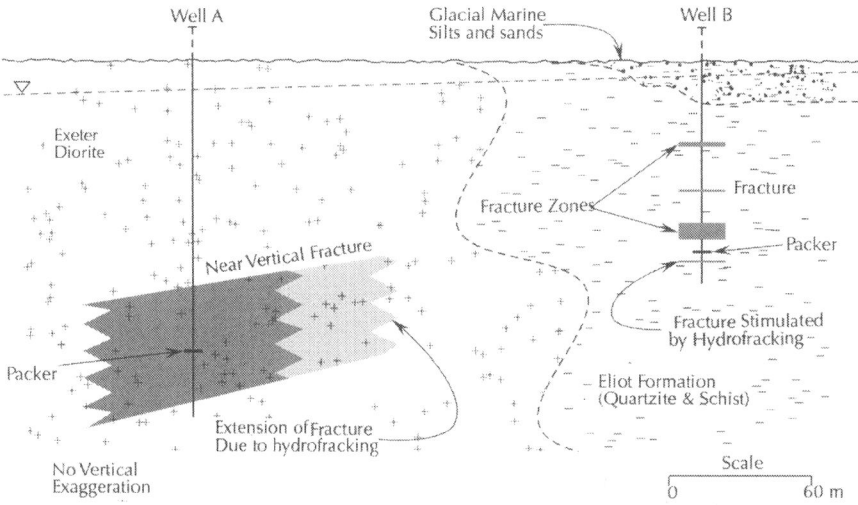

Figure 1: Schematic cross section of the test site depicting the relation of Well A to Well B and preand post-hydraulic fracturing fractures. A near vertical fracture intersected by Well A is shown extending in the direction of Well B after hydraulic fracturing and a fracture stimulated by hydraulic fracturing is shown below the packer in Well B. Note that packers, above ground well casings, and well caps are not to scale; the Eliot Formation is reportedly a schist but the Well B drill log reported mostly quartzite.

Description of the Site Test Wells

Well A was drilled into granitic rock (Exeter Diorite) to a depth of 146 m. An aplite dike was intersected at depths between 110 and 116 m. Well B was drilled through 17 m of mostly glacial marine silt and fine sand, then to a depth of 91 m in schistose rock (Eliot Formation) with quartzite. Some quartzite veins and one aplite vein were intersected between 41 and 46 m followed by 20 meters of quartzite and then more of the schistose rock (Figure 1).

Geophysical Logs

Prior to and after hydraulic fracturing, Wells A and B were subjected to geophysical logging. Geophysical logging was conducted to assist in detecting the existence of naturally occurring subsurface fractures, changes to the existing fractures after hydraulic fracturing, and stimulated fractures resulting from hydraulic fracturing.

- Well a Geophysical Logs. Stewart [10] indicates that significant changes after hydraulic stimulation of parameters measured before fracturing were only noted in the natural gamma and gamma-gamma logs, which occurred at the 137 - 144 m interval, corresponding to the depth of the only water-bearing zone in the well. The depths at which parameters were measured in the geophysical logs did not change after hydraulic fracturing, but there were increases in their magnitudes. The geophysical logs thus suggest that there was an extension of a prominent fracture in Well A at 143 m, but no new fractures were formed after hydraulic fracturing. The gamma and gamma-gamma logs, as well as the neutron radiation logs, also show changes in radiation levels at the depths of an aplite dike (110 - 116 m) that is located above the targeted hydraulic fractured zone.

- Well B Geophysical Logs. Stewart [10] noted many more changes in the geophysical logs for Well B. These seemed to be concentrated around intervals centered on the following depths: 33.5 - 35 m, 66 m, 72 m, and small irregularities at 81 m, which became more numerous and pronounced after hydraulic fracturing. Dewatering of only the single water-bearing fracture zone at 33.5 - 35 m is observed during tests prior to hydraulic fracturing, while dewatering of fractures at 33.5 - 35, 53 and 66 m occur after the hydraulic fracturing procedure. It is noted that the fractures at 53 and 66 - 72 m were located above the depth of the packer, and should not have been formed as a result of the hydraulic fracturing process.

Reinterpretation of Well-Test Data Using Inverse Analysis and Related Methods

As previously stated use of inverse analysis, diagnostic plots, or derivative analysis methods provide for a more reliable diagnostic and quantitative analysis of pumping tests. Application of these methods to the Stewart [10] data can be used as a post-audit on the effect of hydraulic fracturing of the two test wells analyzed by Stewart [10] and subsequently by dos Santos et al. [3].

Some of the flow mechanisms potentially detected in this study are (1) on a log-log radial flow plot, wellbore storage effects (unit slope-early time) and a closed aquifer (unit slope-late time), (2) on a semi-log radial flow plot, infinite acting radial flow or IARF (constant slope at late time), (3) on a log-log linear flow plot, an infinite or high conductivity fracture (unit slope at early time), (4) on a log-log bilinear flow plot, a finite or low conductivity fracture (unit slope at early time), and (5) on a radial flow plot, a constant head boundary (zero slope at late time). One or more changes in the flow regime occurred during several of the tests. In other cases, the response curves for different tests were similar. The diagnostic and derivative plots for the most representative examples are given in this paper.

In the past, semi-log plots were used for diagnosis of which solutions would best fit a set of data. In some unique cases (e.g., well-bore storage, discrete fractures, radial flow, or closed aquifers) specialized plots could be used to identify specific flow regimes. Log-log plots were then used for type curve fitting to determine aquifer constants. Derivatives can now be used in conjunction with semi-log plots to determine which conceptual model best fits a set of data, [8], and the results of automatic type curve fitting can easily be depicted on a semi-log plot. For these reasons and to present the data as concisely as possible, only semi-log graphs are included in this paper. Specialized plots were used to determine such effects as those of well-bore storage, discrete fractures, or a closed aquifer, the results of which are discussed.

Pre-Hydraulic Fracturing Well a Tests

A summary of the single-well aquifer test conducted in well a prior to hydraulic fracturing is given in Table 1. The tests varied in pumping rates and test duration. A summary of the model analysis is given in Table 2.

Table 1: Aquifer test specifics for well a prior to hydraulic fracturing

Test Name	Test Date	Pumping Rate (L·min⁻¹)	Test Duration (min)	Static Water Level (m)	Comment
A2	May 24, 1973	37.85	420	12.3	Aplite dike intersected at 98 - 104 m. Fracture zone intersected at 126 - 131 m. At 230 - 420 min dewatering of major fracture.
A3	May 30, 1973	37.85	380	12.3	From 0 - 50 min no fit was obtained. Probable combination of well-bore storage and linear flow.
A5	June 7, 1973	15.14	860	12.5	At 800 - 860 min dewatering of major fracture.

Table 2: Aquifer test analyses for well a prior to hydraulic fracturing

| Test Name | Analysis Method | | Test Period (min) | Aquifer Paramters | | Source | Figure Number |
	Model Type	Derivative Type		Transmissivity (m²·s⁻¹)	Storage (dimen.)		
A2							
	Papadopulus-Cooper	Well-Bore Storage	0 - 60	6.70 × 10⁻⁷	0.015	This Study	...

	Moench 1 & 3	Well-Bore Storage	0 - 60	1.70×10^{-7}	0.119	This Study	2
	Cooper-Jacob	Leaky	60 - 130	1.95×10^{-6}	0.064	This Study	...
	Hantush A/S	Leaky	60 - 130	8.10×10^{-7}	0.001	This Study	2
	SVF (F)	Leaky	60 - 130	1.89×10^{-6}	4.0×10^{-5}	This Study	...
	Cooper-Jacob	Leaky	130 - 230	3.60×10^{-6}	0.002	This Study	...
	Hantush A/S	Leaky	130 - 230	1.78×10^{-6}	0.001	This Study	2
	Moench 1 & 3	Leaky	130 - 230	2.53×10^{-6}	8.0×10^{-6}	This Study	...
	SVF (F)	Leaky	130 - 230	3.63×10^{-6}	3.0×10^{-6}	This Study	...
A3							
	SVF (I)	Linear	20 - 200	5.40×10^{-7}	0.002	This Study	3
	Moench 1 & 3	Leaky	200 - 380	1.15×10^{-6}	0.432	This Study	3
	Hantush A/S	Leaky	200 - 380	1.07×10^{-6}	0.429	This Study	...
	Cooper-Jacob	Leaky	200 - 380	1.68×10^{-6}	0.423	This Study	...
A5							
	Hantush A/s	Leaky	50 - 500	4.00×10^{-7}	0.208	This Study	4
	Cooper-Jacob	Leaky	50 - 500	6.10×10^{-7}	0.207	This Study	...
	Hantush A/S	Leaky	500 - 800	5.30×10^{-7}	0.086	This Study	...
	Cooper-Jacob	Leaky	500 - 800	9.40×10^{-7}	0.053	This Study	...
	Moench 1 & 3	Leaky	500 - 800	7.30×10^{-7}	0.006	This Study	4
	Papadopulus-Cooper	Leaky	0 - 860	2.46×10^{-6}	7.0×10^{-6}	This Study	...
	Barenblatt	...	...	7.20×10^{-7}	0.13	[10]	...
	Papadopulus-Cooper	...	...	2.41×10^{-6}	9.0×10^{-6}	[3]	...

Note: Hantush A/S = Leaky aquifer with aquitard storage. SVF (F) = Single vertical fracture (Finite conductivity/uniform flux). SVF (I) = Single vertical fracture (infinite conductivity).

- Test A2. The diagnostic and derivative curves, shown on a semi-log plot (Figure 2), from Test A2 indicate that there are several changes in the flow-controlling mechanism, while pumping the well at 37.85 L·min^{-1} (Table 1). Before 60 minutes, both curves exhibit the typical response associated with wellbore storage effects, in which the drawdown and its derivative exhibit unit slopes at early time, and the derivative reached a peak at the end of the well-bore storage period.

The best overall fit was achieved by using the Moench leaky aquifer model (Cases 1 and 3) producing a transmissivity T of 1.70 × 10^{-7} m^2·s^{-1} (Table 2). Case 1 assumes overlying and underlying constant head sources, while Case 3 replaces the underlying source with an impermeable boundary. There is little difference in the fit of the Papadopulos-Cooper model (T = 6.70 × 10^{-7} m^2·s^{-1}) (Table 2) relative to the Moench leaky aquifer model. Both solutions account for well-bore storage effects, but only the Moench solution considers the effects of leakage, so it is considered to be more reliable.

At 60 minutes there is a sharp change in slope at the same depth as the 6-m thick aplite dike when the drawdown equaled 98 m. Between 60 and 150 minutes, the drawdown forms a fairly straight-line segment having drawdowns of 99 and 120 m, respectively. While a dike may cause a deflection of a drawdown curve; in this case, the straight-line segment occurs over a greater depth range than the thickness of the dike. This suggests that the deflection of the curve was probably not due to the presence of the much thinner dike, but more likely represented a change in the flow regime from the effects of well-bore storage to a leaky aquifer. The Hantush leaky aquifer model with aquitard storage best fit the data for that segment, with a T = 8.10 × 10^{-7} m^2·s^{-1} (Table 2). It is noted that the Hantush and Moench Case 1 models both use the same solutions, but the Hantush model does not account for wellbore storage. The Cooper-Jacob IARF model and the Gringarten-Witherspoon single vertical fracture models also provide good fits to the data, both with T values of 1.89×10^{-6} m^2·s^{-1} (Table 2). The sampling period was only 0.3 log units, which may explain the relatively good fit of multiple models because it was shorter than the time period (at least one log unit) generally recommended for type curve analyses. During the early part of the segment, the drawdown forms a straight line on a plot of s versus $t^{1/4}$ (not shown) and a steadily increasing drawdown of the derivative, characteristic of bilinear flow, which occurs when flow is controlled by a finite conductivity vertical fracture. This response is followed by a steady decrease of the drawdown of the derivative, which is typical of a leaky aquifer. The analyses of the derivative and diagnostic plots (Figure 2) suggest that both a finite conductivity vertical fracture and leakage affected groundwater flow during this segment.

At 130 minutes and s = 120 m there is another decline in the slope of the curve that probably represents a second area of higher T, intersected as a result of the expansion of the trough of depression. In this case, the Hantush leaky aquifer (T = 1.78×10^{-6} m^2·s^{-1}), Cooper-Jacob (IARF) (T = 3.60×10^{-6} m^2·s^{-1}) and the Gringarten-Witherspoon single, vertical fracture (T = 3.63×10^{-6} m^2·s^{-1}) models also provide potentially reliable results (Table 2). The Moench (Cases 1 and 3) solutions do not provide as good a fit to the data, but the result (T = 2.53×10^{-6} m^2·s^{-1}) (Table 2) allows for comparison with post-hydraulic fracturing tests. The derivative plot indicates that an initial IARF segment is followed by leakage from 130 minutes to the end of the pumping test, and suggests that the Hantush leaky aquifer model best fit the data. At 230 minutes, a constant drawdown occurred that lasted until the end of the test, with an s = 127 m. This response appeared to be due to dewatering of the fracture zone at s between 126 and 131 m. The derivative indicates that dewatering of the fracture produced the same effect as a constant head boundary, since in both cases the derivative declines to zero drawdown.

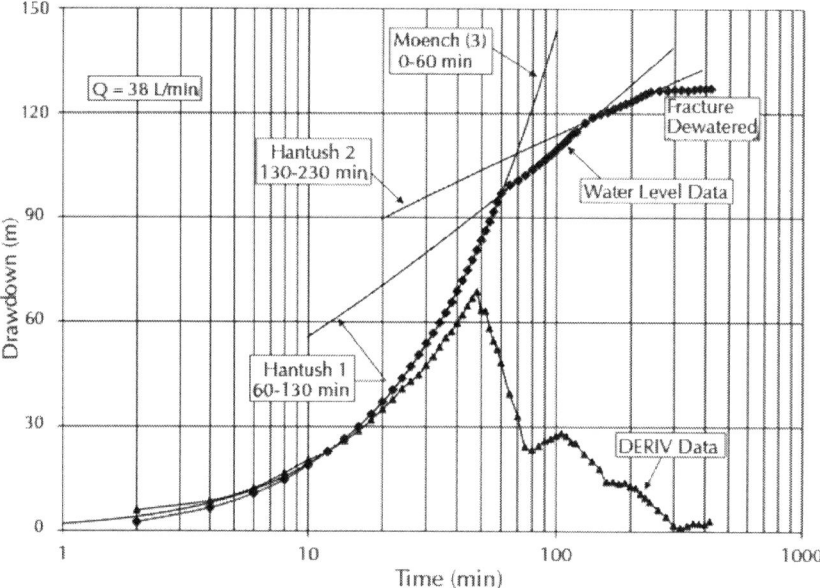

Figure 2: Test A2, semi-log plot of drawdown (Water Level Data) and logarithmic derivative (DERIV Data) for a 38 L·min^{-1}, 420 minutes test with the following flow regimes: Moench 3-wellbore storage (0 - 60 min), Hantush-leaky aquifer (segment) 1 (60 – 130 min), Hantush-leaky aquifer (segment) 2 (130 - 230 min) and fracture dewatered (230 - 420 min).

- Test A3. Well A was pumped at the same rate as Test A2 (37.85 L·min^{-1}) during Test A3 (Table 1), but the rate of drawdown was slower during Test A3. There is no evidence of any effects due to well bore storage during Test A3, while depletion of well bore storage may have produced the greater rate of drawdown observed during Test A2. The Gringarten-Witherspoon model (Figure 3) for a single vertical fracture, with infinite or high conductivity, best fit the data during the period 20 - 200 minutes, after the effects of well-bore storage and before a change in T, which is similar to that noted in Test A2. This produced a calculated T = 5.40 × 10^{-7} m^2·s^{-1} (Table 2). During Test A2 bilinear flow is noted during the second segment of that test, while linear flow occurs in Test A3, indicating that well development may have increased the conductivity of the fracture. Although dos Santos et al. [3] calculated a fracture length, we make no such estimates because models used to analyze discrete fractures require drawdown data from two or more observation wells, and no such data were taken at the New Hampshire test site. From 200 minutes until the end of Test A3, a T = 1.15×10^{-6} m$^2 \cdot$s^{-1} was calculated using the Moench 1 and 3 solutions, with good fits also made using the Hantush and Cooper-Jacob methods (T = 1.07×10^{-6} m$^2 \cdot$s^{-1} and 1.68×10^{-6} m$^2 \cdot$s^{-1}, respectively) (Table 2). These results suggest a trough of depression may have reached the first area of higher T, but not the second area of even higher T noted during Test A2.

- Test A5. Well A was pumped at a rate of 15.14 L·min^{-1} during Test A5 (Table 1). The derivative curve and visual inspection for this test (Figure 4) suggest that leakage dominated the flow regime. The Hantush and Moench leaky aquifer models best fit the data. However, the Hantush and Moench leaky aquifer models required the type curves be fitted to two different segments, 50 - 500 and 500 - 800 minutes (Table 2). No model fit the drawdown during the first 50 minutes, because that segment appears to have been influenced by both well-bore storage and linear flow effects (slope = 0.75). Between 50 and 500 minutes, the Hantush model (Figure 4) produced a T = 4.00×10^{-7} m$^2 \cdot$s^{-1} (Table 2). Between 500 and 800 minutes, the curve flattens due to a slight increase in T, but the Moench 1 and 3 models best fit the data, producing a T = 7.30×10^{-7} m$^2 \cdot$s^{-1} (Table 2). After 800 minutes, the water level stabilizes due to dewatering of the primary water-bearing fracture. These data indicate that, at the lower pumping rate used during this test, the trough of depression did not reach the areas with higher T values.

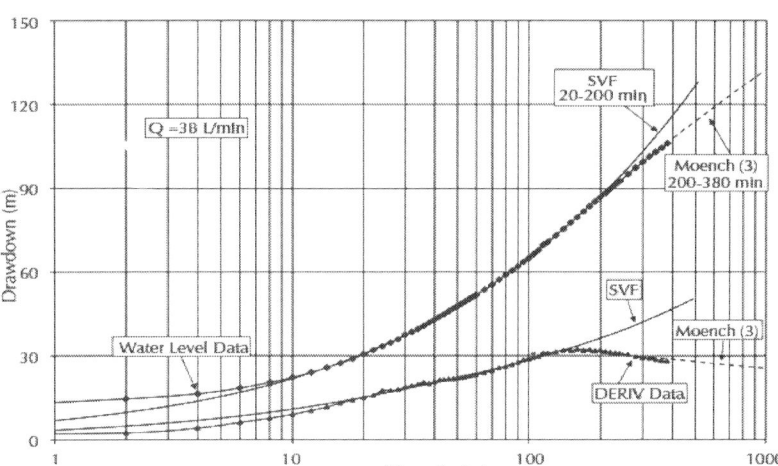

Figure 3. Test A3, semi-log plot of drawdown (Water Level Data) and logarithmic derivative (DERIV Data) for a 38 L·min^{-1}, 380 minutes test with the following flow regimes: SVF(1) (20 - 200 min), Moench 3-leaky aquifer (200 - 380 min).

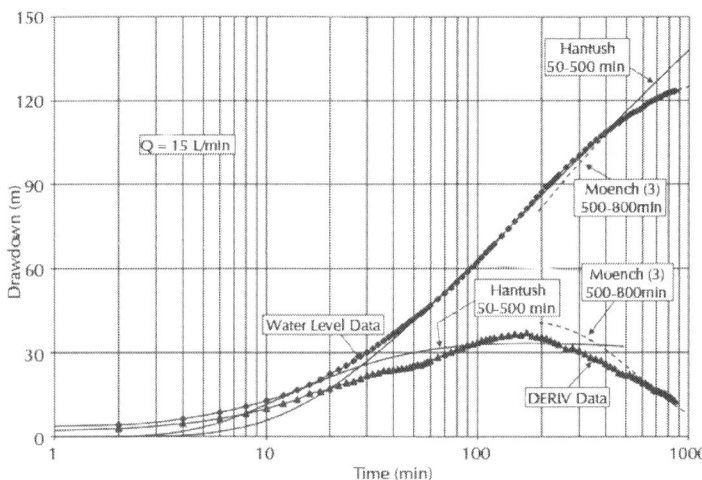

Figure 4. Test A5, semi-log plot of drawdown (Water Level Data) and logarithmic derivative (DERIV Data) for a 15 L·min^{-1}, 860 minutes test with the following flow regimes: Hantush-leaky aquifer (50 - 500 min), Moench 3-leaky aquifer (500 - 800 min).

Pre-Hydraulic Fracturing Well B Tests

A summary of the single-well tests conducted in Well B is given in Table 3. Two tests were conducted at different pumping rates and duration periods.

- Tests B3 and B4. Well B was pumped at 18.925 L min^{-1} and 15.14 L·min^{-1} during Tests B3 and B4, respectively (Table 3). Similar responses occurred during both Tests B3 (Figure 5) and B4. There are early leaky aquifer responses that lasted until the drawdown reaches 30 m, where a single, discrete, water-bearing fracture is intersected by Well B. At that point there are rapid drawdowns in the water levels and derivatives, due to dewatering of the fracture. There is no doubling of the slopes, which is expected for the case where barriers are present, as proposed by both Stewart [10] and dos Santos et al. [3]. The late-time unit slopes of the drawdowns are indicative of a closed aquifer.

The Moench model (Cases 1 and 3) provided a good overall fit to the early-time data of both tests. The higher T (6.38 × 10^{-6} m^2·s^{-1}) calculated for Test B4 relative to Test B3 (3.25 × 10^{-6} m^2·s^{-1}) may reflect the effects of well development (Table 4). The Papadopulos-Cooper model also provided a good fit to the data, but neither drawdown nor derivative plots have the characteristic unit slope associated with the effects of well bore storage or the peak in the derivative that occurs after the storage is depleted. Application of the of Gringarten and Ramey [48], used by dos Santos et al. [3], to the early-time data from the two tests, yields calculated T values of 1.70 × 10^{-7} m^2·s^{-1} and 4.96 × 10^{-5} m^2·s^{-1} (Table 4) for Tests B3 and B4, respectively.

Attempts were made to match the late-time data to all models, but good fits could not be obtained. However, the results did indicate that the T values were low (2.3 × 10^{-7} and 2.4 × 10^{-7} m^2·s^{-1}) as were those of (3.70 × 10^{-8} m^2·s^{-1} and 9.30 × 10^{-8} m^2·s^{-1}) obtained by dos Santos et al. [3], suggesting that those authors used late-time data in their calculations, when closed aquifer conditions were present. None of the data fit either the Moench or Barker double-porosity models since there are no sigmoidal shapes of the drawdown curves or depressions of the derivative curves associated with the dual porosity behavior assumed by Stewart [10].

Post-Hydraulic Fracturing Aquifer Test Analyses

Post-Hydraulic Fracturing Well a Tests

Several tests were conducted in well a after hydraulic fracturing (Table 5). The tests varied in pumping rates and duration. Prior to stimulation, a packer was placed in Well A between the aplite dike at 110 - 116 m and the prominent, vertical, water-bearing fracture at 143 m. As a result of the hydraulic fracturing procedure, the geophysical logs indicate that there was an extension of that fracture, but no new fractures were formed in the well.

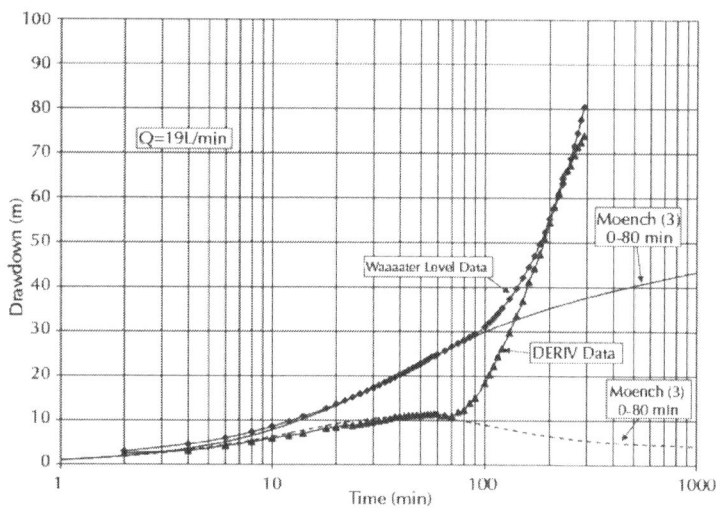

Figure 5: Test B3, semi-log plot of drawdown (Water Level Data) and logarithmic derivative (DERIV Data) for a 19 L·min^{-1}, 290 minutes test with the following flow regimes: Moench 3-wellbore storage and leaky aquifer (0 - 80 min), to closed aquifer (80 - 290 min).

Table 3: Aquifer test specifics for Well B prior to hydraulic fracturing

Test Name	Test Date	Pumping Rate (L min⁻¹)	Test Duration (min)	Static Water Level (m)	Comment
B3	June 19, 1973	18.925	290	6.2	Main fracture intersected at 29 m.
B4	June 20, 1973	15.14	460	6.6	Main fracture intersected at 28.5 m.

- Test A14. Well A was pumped at a rate of 94.625 L·min⁻¹ for 400 minutes after hydraulic fracturing during Test A14 (Table 5). Initially, when evaluating the data from Test A14, the best results were achieved when applying the Moench 1 and 3 models to the full set of data (0 - 400 min), with a T value of 1.09×10^{-6} m²·s⁻¹ (Table 6). It appears, however, that the model deviates from the data on semi-log diagnostic and derivative plots at late time (~250 min; s = 101 m) (Figure 6). Similar deviations are more apparent during the three other tests (Tests A16, A741, A742) of the well (130 - 230 min; s = 65 - 87 m). In all cases, convergence between the model and data occurred using the full set of data. Better fits are achieved when analyses using the Hantush solution excluded the apparent late-time fluctuations in the water levels, as well as the early-time wellbore storage effects.

- Using the drawdown data from Test A14 and a window of 0 - 30 minutes (Figure 6) the Moench 1 and 3 and Papadopulos-Cooper models both produced good results and the same T value (8.29×10^{-6} m²·s⁻¹), while the Moench 3 model, when applied to the 30 - 200 minutes period, achieved the best result (T = 4.08×10^{-6} m²·s⁻¹) (Table 6). These results indicate higher T values occurred after hydraulic fracturing (Table 6) than before hydraulic fracturing (Table 2) of Well A. The late-time decline in T during Test A14 is attributed to dewatering of a shallow aquifer, possibly formed by the sandy portion of the surficial material overlying the bedrock near Well B.

- Test A16. Well A was pumped at 75.7 L min^{-1} for 360 minutes during Test A16 after hydraulic fracturing (Table 5). Using the drawdown data from Test A16 and a window of 0 - 100 minutes, the Moench 1 and 3 (Figure 7) and Papadopulos-Cooper models both produced good results, but different T values (3.16 × 10^{-6} m^2·s^{-1} and 1.10 × 10^{-5} m^2·s^{-1}, respectively) (Table 6). The derivative indicates that the response is typical of a leaky aquifer flow regime and that there is no evidence of wellbore storage effects. This suggests that the Moench model best described the flow regime. The lower value of T relative to Test A14 is more consistent with continued dewatering of a permeable zone.
- Test A741. Well A was pumped at 75.7 L·min^{-1} for 600 minutes during Test A741 about one year after hydraulic fracturing (Table 5). The data collected during Test A741 appear to have been erratic, as evidenced by the relatively poor fits of the various models to that data. The best results were achieved using the Moench model (T = 1.55 × 10^{-5} m^2·s^{-1}) (Figure 8) (Table 6). Test A742. Similar to Test A14, Well A was pumped at 94.625 L·min^{-1} for 600 minutes during Test A742 after hydraulic fracturing (Table 5). The data collected during Test A742 show somewhat erratic behavior, but fairly good fits were achieved when the various models were applied. The best result is achieved using the Hantush model (T = 3.30 × 10^{-6} m^2·s^{-1}), but fairly good fits were also obtained using the Moench model (T = 7.00 × 10^{-6} m^2·s^{-1}), and the Papadopulos-Cooper models (T = 1.00 × 10^{-5} m^2·s^{-1}) (Figure 9) (Table 6). The derivative indicates a response typical of a leaky aquifer flow regime with no wellbore storage effects. The lower T values relative to Test A741 provided additional evidence for of the effects of dewatering of a permeable zone. The erratic behavior noted during tests A741 and A742 is attributed to errors in transcribing drawdown data from the log-log graphs in the Stewart [10] paper to the speadsheets used to generate plots in the present study.

Table 4: Aquifer test analyses for Well B prior to hydraulic fracturing

Test Name	Analysis Method		Test Period (min)	Aquifer Paramters		Source	Figure Number
	Model Type	Derivative Type		Transmissivity ($m^2 \cdot s^{-1}$)	Storage (dimen.)		
B3							
	Moench 1 & 3	Well-Bore Storage & Leaky	0 - 80	3.25×10^{-6}	0.015	This Study	5
	Hantush A/S	Well-Bore Storage & Leaky	0 - 80	2.49×10^{-6}	0.119	This Study	...
	Papadopulus-Cooper	Well-Bore Storage & Leaky	0 - 80	6.81×10^{-6}	0.064	This Study	...
	SHF	Well-Bore Storage & Leaky	0 - 80	1.70×10^{-7}	0.001	This Study	...
	Moench Double Porosity	Well-Bore Storage & Leaky	0 - 80	3.38×10^{-6}	4.0×10^{-5}	This Study	...
	SHF	...	100 - 290	2.30×10^{-7}	0.002	This Study	...
	Barenblatt	...	...	2.87×10^{-6}	0.001	[10]	...
	SHF	...	...	3.70×10^{-8}	8.0×10^{-6}	[3]	...
B4							
	Moench 1 & 3	Well-Bore Storage & Leaky	0 - 160	6.38×10^{-6}	0.002	This Study	...
	Hantush A/S	Well-Bore Storage & Leaky	0 - 160	2.42×10^{-6}	0.321	This Study	...
	Papadopulus-Cooper	Well-Bore Storage & Leaky	0 - 160	6.38×10^{-6}	0.002	This Study	...
	SHF	Well-Bore Storage & Leaky	0 - 160	4.96×10^{-5}	2.0×10^{-4}	This Study	...
	Moench Double Porosity	Well-Bore Storage & Leaky	0 - 160	6.39×10^{-6}	2.0×10^{-5}	This Study	...
	SHF	Closed Aquifer	200 - 460	2.40×10^{-7}	0.051	This Study	...
	Barenblatt	...	...	3.02×10^{-6}	0.190	[10]	...
	SHF	...	...	9.30×10^{-8}	0.159	[3]	...

Note: Hantush A/S = Leaky aquifer with aquitard storage. SHF = Single horizontal fracture.

Table 5: Aquifer test specifics for well a after hydraulic fracturing

Test Name	Test Date	Pumping Rate (L·min^{-1})	Test Duration (min)	Static Water Level (m)	Comment
A14	July 10, 1973	94.625	400	12.6	
A16	July 13, 1973	75.7	360	12.6	
A741	August 15, 1974	75.7	600	...	Static water level not reported.
A742	August 16, 1974	94.625	600	...	Static water level not reported.

Post-Hydraulic Fracturing Well B Tests

Stewart [10] reported a single producing fracture zone in Well B at 32 - 35 m and changes to the geophysical logs' parameters at 66 and 72 m before and stimulation of the irregularities at 82 m after hydraulic fracturing. The packer was set at 76 m prior to hydraulic fracturing, so occurrence of a stimulated fracture was significant. Several single-well tests were conducted in Well B after hydraulic fracturing (Table 7).

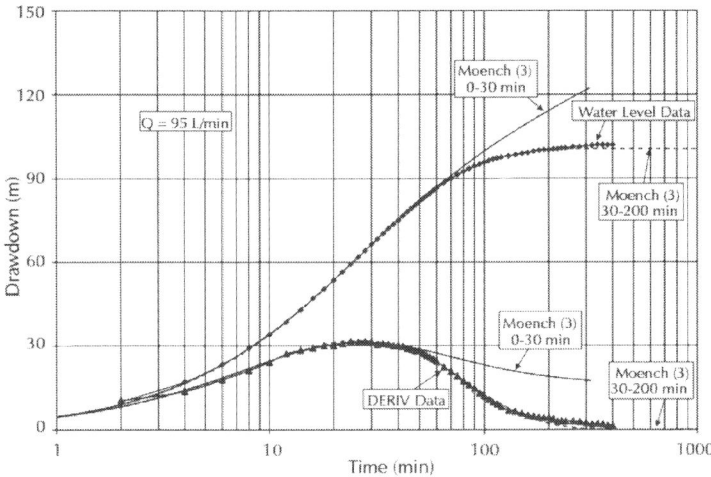

Figure 6: Test A14, semi-log plot of drawdown (Water Level Data) and logarithmic derivative (DERIV Data) for a 95 L·min^{-1}, 400 minutes test with the following flow regimes: Moench 3-leaky aquifer (0 - 30 min) and Moench 3-leaky aquifer (30 - 200 min).

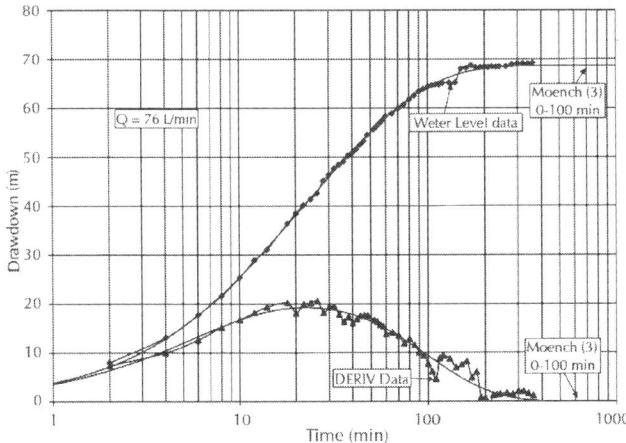

Figure 7: Test A16, semi-log plot of drawdown (Water Level Data) and logarithmic derivative (DERIV Data) for a 76 L·min^{-1}, 360 minutes test with the following flow regimes: Moench 3-leaky aquifer (0 - 100 min).

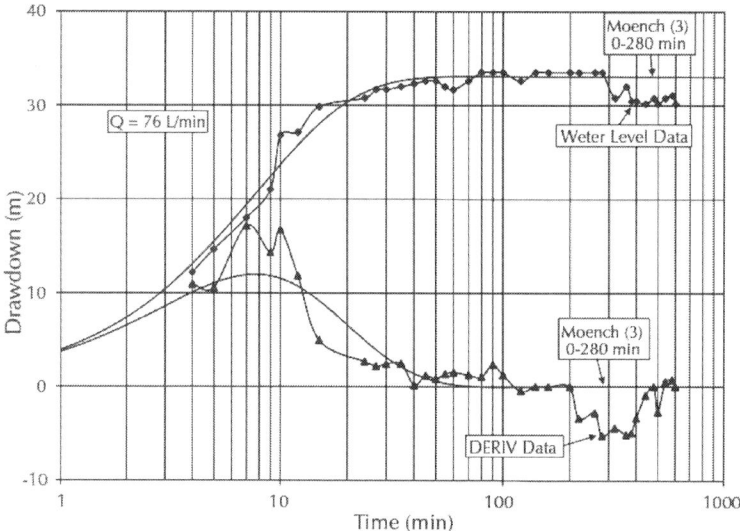

Figure 8: Test A741, semi-log plot of drawdown (Water Level Data) and logarithmic derivative (DERIV Data) for a 76 L·min^{-1}, 600 minutes test with the following flow regimes: Moench 3-leaky aquifer (0 - 280 min).

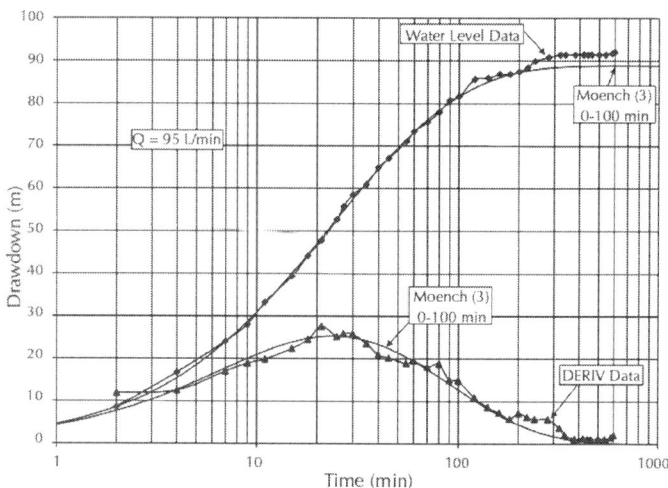

Figure 9: Test A742, semi-log plot of drawdown (Water Level Data) and logarithmic derivative (DERIV Data) for a 95 L·min^{-1}, 600 minutes test with the following flow regimes: Moench 3-leaky aquifer (0 - 100 min).

- Test B25. Well B was pumped at 38.85 L·min⁻¹ for 360 minutes (Table 7) during Test B25 after hydraulic fracturing. Drawdowns and derivatives show a leaky aquifer response for the first 60 minutes with the Moench model (Cases 1 and 3) providing a good fit to the drawdown data (T = 6.71 × 10⁻⁶ m²·s⁻¹) (Figure 10) (Table 8). The Papadopulos-Cooper model (T = 1.87 × 10⁻⁵ m²·s⁻¹) (Figure 10) (Table 8) also provides a good fit, but there is no characteristic drawdown and derivative unit slopes associated with well bore storage effects, and the derivative indicated that a leaky aquifer flow regime was present.

Table 6: Aquifer test analyses for Well A after to hydraulic fracturing

Test Name	Analysis Method		Test Period (min)	Aquifer Paramters		Source	Figure Number
	Model Type	Derivative Type		Transmissivity ($m^2 \cdot s^{-1}$)	Storage (dimen.)		
A14							
	Papadopulus-Cooper	Leaky	0 - 30	8.29×10^{-6}	0.016	This Study	6
	Moench 1 &3	Leaky	0 - 30	8.29×10^{-6}	0.016	This Study	...
	Hantush A/S	Leaky	10 - 150	3.00×10^{-6}	0.190	This Study	6
	Moench 1 & 3	Leaky	30 - 200	4.08×10^{-6}	0.061	This Study	...
	Hantush A/S	Leaky	0 - 400	2.60×10^{-6}	0.346	This Study	...
	Moench 1 & 3	Leaky	0 - 400	1.09×10^{-6}	0.007	This Study	...
	Papadopulus-Cooper	Leaky	0 - 400	2.25×10^{-5}	2.0×10^{-6}	This Study	...
	Barenblatt	...	...	1.21×10^{-5}	0.004	[10]	...
	Papadopulus-Cooper	...	...	2.43×10^{-5}	5.0×10^{-7}	[3]	...
A16							

	Papadopulus-Cooper	Leaky	0 - 100	1.10×10^{-5}	0.014	This Study	...
	Hantush A/S	Leaky	10 - 100	1.51×10^{-6}	0.059	This Study	...
	Moench 1 & 3	Leaky	0 - 100	3.16×10^{-6}	0.019	This Study	7
	Moench Double Porosity	Leaky	0 - 100	6.31×10^{-6}	2.0×10^{-4}	This Study	...
	Hantush A/S	Leaky	0 - 360	2.05×10^{-6}	0.134	This Study	...
	Moench 1 & 3	Leaky	0 - 360	2.83×10^{-6}	0.032	This Study	...
	Papadopulus-Cooper	Leaky	0 - 360	1.60×10^{-5}	0.001	This Study	...
	Barenblatt	...	...	1.44×10^{-5}	0.004	[10]	...
	Papadopulus-Cooper	...	...	1.62×10^{-5}	0.001	[3]	...
A741							
	Hantush A/S	Leaky	0 - 280	7.50×10^{-7}	0.091	This Study	...
	Moench 1 & 3	Leaky	0 - 280	1.55×10^{-5}	9.0×10^{-6}	This Study	8
	Papadopulus-Cooper	Leaky	0 - 600	1.55×10^{-5}	2.0×10^{-10}	This Study	...
	Barenblatt	...	...	2.75×10^{-5}	0.003	[10]	...
	Papadopulus-Cooper	...	...	6.37×10^{-5}	4.0×10^{-7}	[3]	...
A742							
	Papadopulus-Cooper	Leaky	0 - 100	1.00×10^{-5}	0.022	This Study	...
	Hantush A/S	Leaky	10 - 100	3.30×10^{-6}	0.384	This Study	...
	Moench 1 & 3	Leaky	0 - 100	7.00×10^{-6}	0.062	This Study	9
	Moench Double Porosity	Leaky	0 - 100	7.00×10^{-6}	3.0×10^{-4}	This Study	...
	Hantuch A/S	Leaky	0 - 600	3.60×10^{-6}	0.336	This Study	...
	Moench 1	Leaky	0 - 600	8.03×10^{-6}	0.046	This Study	...
	Moench 3	Leaky	0 - 600	8.01×10^{-6}	0.046	This Study	...
	Papadopulus-Cooper	Leaky	0 - 600	1.60×10^{-5}	0.001	This Study	...
	Barenblatt	...	...	1.71×10^{-5}	0.001	[10]	...
	Papadopulus-Cooper	...	...	1.72×10^{-5}	6.0×10^{-4}	[3]	...

Note: Hantush A/S = Leaky aquifer with aquitard storage.

Table 7: Aquifer test specifics for Well B after hydraulic fracturing

Test Name	Test Date	Pumping Rate (L·min⁻¹)	Test Duration (min)	Static Water Level (m)	Comment
B25	July 19, 1973	37.85	360	10.16	Fractured dewatered a1 22 - 25 m.
B27	July 23, 1973	56.775	400	9.9	New fractures at 40?, 56 - 62, and 71 m.
B28	July 24, 1973	56.775	260	10.3	
B743	July 30, 1974	37.85	340	...	Static water level not reported.
A744	August 10, 1974	37.85	910	...	Static water level not reported.

After 60 minutes, dewatering of fractures at $s = 22$ and 25 m controls the flow regime for the remainder of the test. At the first fracture, the water level stabilizes somewhat for about 40 minutes, after which there is a sharp decline until the water level reaches the next fracture. The water level subsequently stabilizes at the second fracture for the remainder of the test. The derivative forms a sharp peak, probably due to a rapid decline related to dewatering of the first fracture followed by recovery due to leakage. The derivative drops to zero and remained at that level while the deeper fracture is dewatered, suggesting that the fracture at $s = 25$ m has a higher capacity than the one at $s = 22$ m. Stewart [10] also considered the deeper fracture to be the primary water-bearing fracture in the well. Stewart [10] also stated that the geophysical logs showed no clear evidence that new fractures were formed after hydraulic fracturing. Although the irregularities at 81 m were stimulated, there is no evidence that they were then water-bearing or increased the yield of the well. The relatively small change in T after the hydraulic fracturing process was complete would appear to confirm the Stewart [10] observation.

There may have been natural fractures opened as a result of increased well development due to the extensive pumping (36 h) required to flush the proppant sand from the bottom of the well and to conduct Test B25. Neither optical logs were available nor were packer tests conducted that could provide information on the orientation and hydraulic characteristics of the fractures.

- Test B27. At the pumping rate of 56.775 L·min^{-1} used during Test B27 (Table 7), a leaky aquifer response is obtained during the first 20 minutes. The drawdown data are best fitted by the Hantush model, yielding a T = 1.62 × 10^{-6} m^2·s^{-1} (Table 8), but a good fit was also achieved when the Moench model was applied to the data (T = 2.37 × 10^{-6} m^{-1}·s^{-1}) (Figure 11). In both cases, the T values are about one third of those derived during Test B25. This could have resulted from dewatering of a permeable zone.

After 20 minutes, the derivative forms a sharp peak during the dewatering of the first fracture at s = 25 m, followed by a second peak caused by dewatering of a second fracture at s = 40 m. In this instance neither fracture is clearly evident in the drawdown data, especially on the Stewart [10] log-log plots. Finally, there was a sharp increase in the drawdown of the derivative starting at about 150 minutes (s = 51 m). The geophysical logs indicate that there is a fracture zone centered on s = 56 m and s = 62 m, which may explain some of the response. Another possibility is that a fluctuation in the pumping rate may have caused the earlier than expected change in the derivative. The pumping rate was not sustainable, as the well was nearly dry by the end of the test.

- Test B28. At the 56.775 L·min^{-1} pumping rate during Test B28 (Table 7), there was a leaky aquifer response that lasted for the first 16 minutes with the drawdown data best fitting the Moench Case 3 model, producing a T = 5.42 × 10^{-6} m^2·s^{-1} (Table 8). This was somewhat less than the T = 6.71 × 10^{-6} m^2·s^{-1} derived from the B25 test data, but the increase in T relative to that of Test B27 is possibly due to expansion of the trough of depression to an area with a higher T.

After 16 minutes, the derivative formed a sharp peak during dewatering of the first fracture s = 25 m, followed by a second peak caused by dewatering of the second fracture at s = 40 m. Again, the presence of neither fracture is clearly evident in the drawdown data. Finally, there was a sharp decline in the derivative starting, again, at about 150 minutes, with an s = 56 m, followed by a very sharp decline in both the drawdown and derivative curves at 62 m. These occurred at the same depths as the fracture zones shown on the geophysical logs.

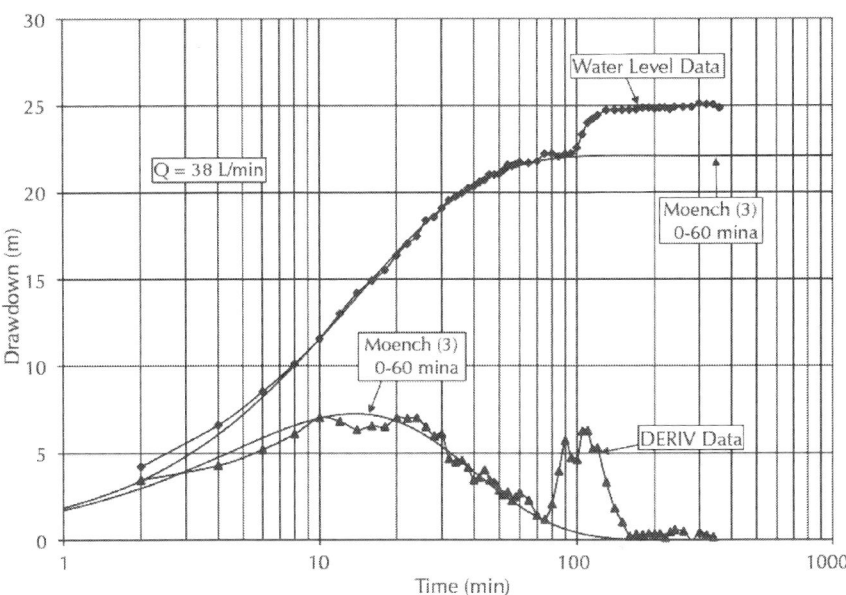

Figure 10: Test B25, semi-log plot of drawdown (Water Level Data) and logarithmic derivative (DERIV Data) for a 38 L·min^{-1}, 360 minutes test with the following flow regimes: Moench 3-leaky aquifer (0 - 60 min).

- Tests B743 and B744. Well B was pumped at 37.85 L·min^{-1} for both Test B743 and Test B744, but Test B743 only lasted for 340 minutes while Test B744 lasted for 910 minutes (Table 7). For Tests B743 and B744, drawdowns appear to have been only sufficient to dewater the shallow fractures at s = 20 and 25 m. There is a peak in the derivative starting at 70 minutes during Test B743 (Figure 12). However, there is no clear evidence of a peak during Test B744.

For Test B743, the best fit to the data was achieved using the Moench leaky aquifer model (Cases 1 and 3), producing a T = 1.49 × 10^{-5} m^2·s^{-1} (Table 8). As with Test A741, one explanation for a higher T value than those noted one year earlier is that long-term pumping of the well caused the trough of depression to extend into an area with an even higher T. This seems possible because the climate was drier, and the water table and stream flow were lower during the test periods in 1974 than in 1973, which would have tended to result in a lower T. Necessary information about the post-hydraulic fracturing operational history of Well B is not presented in either Stewart [10] or dos Santos et al. [3].

The single horizontal fracture model best fit the drawdown data from Test B744. However, the calculated T value (1.08 × 10^{-4} m^2·s^{-1}) (Table 8) is unreasonable, when compared to those obtained from the other tests. The best result appears to be the T = 7.26 × 10^{-6} m^2·s^{-1} obtained using the Hantush leaky aquifer model, although the data also fit the Moench model (T = 6.72 × 10^{-6} m^2·s^{-1}) (Table 8). The decrease in T relative to Test B743 may represent dewatering of a permeable zone.

The Papadopulos-Cooper model also provided a good fit to the post-hydraulic fracturing aquifer test data, but neither the diagnostic nor the derivative plots have the characteristic unit slopes associated with the effects of well bore storage. Also, the characteristic peak for well bore storage is not evident on any of the derivative plots.

The T values calculated from the Well B test data using the SHF model do not provide consistent and realistic values for various reasons. Before hydraulic fracturing the difference in the residuals between the Moench and SHF solutions favors the leaky aquifer model (Test B3) or is insignificant (Test B4). After stimulation, the residuals favored the Moench model, except for Test B744. The derivative of all of the tests, however, is typical of a leaky aquifer response. When the SHF solution is applied to the data, the result is that T values increase more than two orders of magnitude for the data obtained prior to hydraulic fracturing (T = 1.70 × 10^{-7} for Test B3 to 4.96 × 10^{-5} m^2·s^{-1} for Test B4 (Table 4)) and decline by more than two orders of magnitude immediately after stimulation (from T = 3.03 × 10^{-4} for Test B25 to 5.90 × 10^{-7} m$^2 \cdot$s^{-1} for Test B28 (Table 8)). Unfortunately, the results of the dos Santos et al. [3] paper could not be reproduced in the present study. Finally, the best evidence that the Moench leaky aquifer better describes the flow regime is that the T values derived from the Well B tests after hydraulic fracturing when the two wells were hydraulically connected, are nearly identical to those of the Well A tests (ratios of 1:1.0 - 1.3) and the SHF model does not fit any of the Well A test data.

Table 8: Aquifer test analyses for Well B after to hydraulic fracturing

Test Name	Analysis Method		Test Period (min)	Aquifer Paramters		Source	Figure Number
	Model Type	Derivative Type		Transmissivity (m$^2 \cdot$s^{-1})	Storage (dimen.)		
B25							
	Moench 1 & 3	Leaky	0 - 60	6.71 × 10^{-6}	0.038	This Study	10
	Hantush A/S	Leaky	0 - 60	4.75 × 10^{-6}	0.267	This Study	...
	Papadopulus-Cooper	Leaky	0 - 60	1.87 × 10^{-5}	0.004	This Study	...
	SHF	Leaky	0 - 60	3.03 × 10^{-4}	3.0 × 10^{-4}	This Study	...
	Moench Double Porosity	Leaky	0 - 60	7.91 × 10^{-6}	3.0 × 10^{-4}	This Study	...
	SHF	Well-Bore Storage & Leaky	0 - 360	4.52 × 10^{-4}	9.0 × 10^{-4}	This Study	...
	Barenblatt	...	...	2.20 × 10^{-6}	0.034	[10]	...
	SHF	...	...	1.24 × 10^{-5}	1.0 × 10^{-4}	[3]	...
B27							
	Hantush A/S	Leaky	0 - 20	1.62 × 10^{-6}	0.002	This Study	...
	Moench 1 & 3	Leaky	0 - 20	2.37 × 10^{-6}	0.006	This Study	11
	Papadopulus-Cooper	Leaky	0 - 20	1.14 × 10^{-5}	0.049	This Study	...
	Moench Double Porosity	Leaky	0 - 20	4.02 × 10^{-6}	7.0 × 10^{-4}	This Study	...
	SHF	Leaky	0 - 20	3.35 × 10^{-6}	2.0 × 10^{-4}	This Study	...
B28							
	Hantush A/S	Leaky	0 - 16	5.70 × 10^{-6}	0.286	This Study	...
	Moench 1	Leaky	0 - 16	6.57 × 10^{-6}	0.093	This Study	...

	Method	Type	Range	Value 1	Value 2	Source	
	Moench 3	Leaky	0 - 16	5.42×10^{-6}	0.092	This Study	...
	Papadopulus-Cooper	Leaky	0 - 16	9.71×10^{-6}	0.054	This Study	...
	Moench Double Porosity	Leaky	0 - 16	4.49×10^{-6}	6.0×10^{-4}	This Study	...
	SHF	Leaky	0 - 16	5.90×10^{-7}	0.003		
B743							
	Hantush A/S	Leaky	0 - 55	6.64×10^{-6}	0.372	This Study	...
	Moench 1 & 3	Leaky	0 - 55	1.49×10^{-5}	0.011	This Study	12
	Papadopulus-Cooper	Leaky	0 - 55	1.48×10^{-5}	0.011	This Study	...
	SHF	Leaky	0 - 55	8.28×10^{-5}	4.0×10^{-5}	This Study	...
	Moench Double Porosity	Leaky	0 - 55	1.57×10^{-5}	7.0×10^{-5}	This Study	...
	SHF	Well-Bore Storage & Leaky	0 - 340	4.60×10^{-5}	5.0×10^{-6}	This Study	...
	Barenblatt	...	...	1.49×10^{-5}	0.031	[10]	...
	SHF	...	...	6.72×10^{-6}	0.400	[3]	...
B744							
	Moench 1 & 3	Leaky	0 - 60	6.72×10^{-6}	0.133	This Study	...
	Hantush A/S	Leaky	0 - 60	7.26×10^{-6}	0.332	This Study	...
	Papadopulus-Cooper	Leaky	0 - 60	1.38×10^{-5}	0.018	This Study	...
	SHF	Leaky	0 - 60	1.08×10^{-4}	7.0×10^{-5}	This Study	...
	Moench Double Porosity	Leaky	0 - 60	6.93×10^{-6}	1.0×10^{-5}	This Study	...
	SHF	Well-Bore Storage & Leaky	0 - 910	1.18×10^{-3}	0.071	This Study	...
	Barenblatt	...	...	1.22×10^{-5}	0.025	[10]	...
	SHF	...	...	1.70×10^{-5}	1.0×10^{-1}	[3]	...

Note: Hantush A/S = Leaky aquifer with aquitard storage. SHF = Single horizontal fracture.

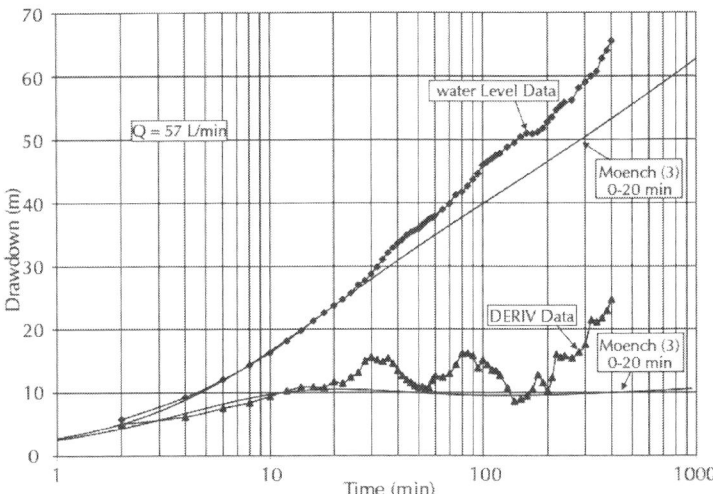

Figure 11: Test B27, semi-log plot of drawdown (Water Level Data) and logarithmic derivative (DERIV Data) for a 57 L·min⁻¹, 400 minutes test with the following flow regimes: Moench 3-leaky aquifer (0 - 20 min).

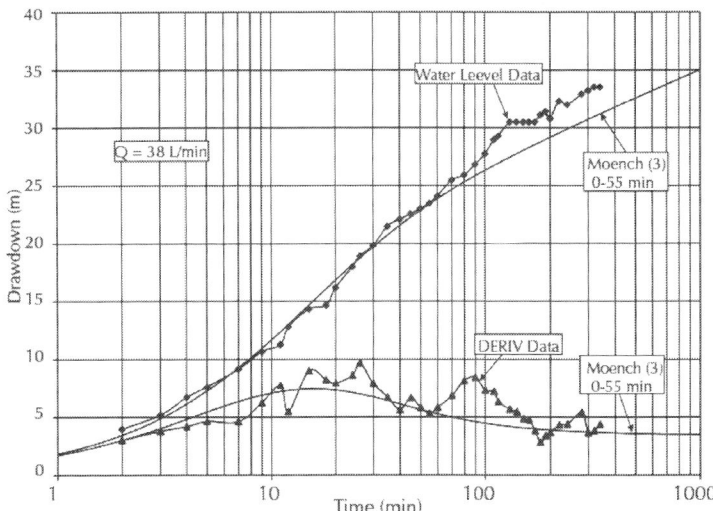

Figure 12: Test B743, semi-log plot of drawdown (Water Level Data) and logarithmic derivative (DERIV Data) for a 38 L·min⁻¹, 340 minutes test with the following flow regimes: Moench 3-leaky aquifer (0 - 55 min).

DISCUSSION

The sequence and order of the changes for the flow-controlling mechanism during the three pre-hydraulic fracturing tests of Well A progressed from early-time well-bore storage (first test, Test A2), through earlyto mid-time single, vertical, extended fracture (second test, Test A3), to midto late-time leaky aquifer effects (Tests A2, A3, and A5). These changes and the associated increasing T values at later times during the individual tests are probably due to the effects of well development as a result of the extensive pumping of Well A, prior to the hydraulic fracturing procedure, and the expansion of the trough of depression to an area with a higher T value. Due to the different pumping rates, analysis periods and models providing best fits to data, it is somewhat difficult to compare the changes between the three tests, but it appears that T progressively declines from one test to the next. This trend was most likely due to dewatering of the aquifer.

The Moench (Cases 1 and 3) model best fits the data from the two tests conducted immediately after hydraulic fracturing of Well A, and from the two tests of that well performed a year later (Table 9). During early time of Test A2, the drawdown and derivative plots have unit slopes typical of well bore storage effects, while leakage occurs at later times. The main advantage of the Moench model is that it considers both well-bore storage and leakage effects, while the Hantush solution neglects the effects of casing storage, which increases early drawdown, and the Papadopulos-Cooper model ignores leakage, which stabilizes late drawdowns. This may be the reason that, relative to the Moench model, the Hantush model under-estimates T values, while the Papadopulos-Cooper solution over-estimates them. The Hantush and Moench Case 1 models both assume upper and lower constant head boundaries, while the Moench Case 3 model assumes constant upper and impermeable lower boundaries. The 16 m of marine silts and clays and 6 m of weathered bedrock, consisting primarily of fine sand, found in Well B form a water table aquifer that could be the source of an upper constant head boundary. Although there are no obvious lower constant head boundaries, the discrete fractures in each well could mimic a weak constant head, so that good fits could be achieved assuming either constant head or impermeable boundaries. The weathered zone exists under water table conditions and is hydraulically coupled to locally confined bedrock. Evidence of this connection is that water levels in wells completed in bedrock follow topography and their yields vary seasonally in response to changing climatic conditions. As a result, leakage from the weathered zone to bedrock can occur, as well as dewatering of the aquifer; both factors which can affect the results of pumping tests.

Dos Santos et al. [3] stated that horizontal fractures formed as a result of the hydraulic fracturing process and based their comment on a study by Murdoch and Slack [49] , which did not include the New Hampshire test site. Murdoch and Slack [49] stated that the hydraulic fractures they created were formed at shallow depths (1 - 12 m) in fine-grained sediments, nearly all of which were in glacial till or weathered formations. At various sites in the Murdoch and Slack [49] study, vertical fractures were formed as depth increased or in relatively massive sediments. Murdoch and Slack did not include fractures formed in competent bedrock at the depths described by Stewart [10] (35 - 130 m). Also, Stewart [10] indicated that most of the natural joints (fractures) in the bedrock (Exeter diorite) are steeply dipping. In some places the fractures may have gentle dips, but in the vicinity of the well sites most were inclined at steep angles.

With the exception of Test A741, the results of the dos Santos et al. [3] study could be reproduced using the Papadopulos-Cooper model, but the residual statistics suggest that there are poor fits between the observed and simulated drawdowns. For Test A741, the T value (1.55×10^{-5} m^2·s^{-1}) obtained using the Moench (Case 1 & 3) solution in the present study is more consistent with the other post-hydraulic fracturing tests using the Papadopulos-Cooper model (T = 1.60×10^{-5} to 2.25×10^{-5} m^2·s^{-1}) than that of dos Santos et al. [3] (T = 6.37×10^{-5} m^2·s^{-1}) (Table 9).

Manual fitting of Barenblatt type curves for a double-porosity aquifer were used in the Stewart [10] study. Our results indicate that none of the data from the tests could fit either of the double-porosity models of Moench [23] or Barker [25]. In addition, analytical plots and derivative analyses indicated that the sigmoidal shape of the drawdown curves or depressions in the derivative curves typical of a double-porosity aquifer were not present during any of the tests.

Prior to hydraulic fracturing, the late-time T during Test A2 was 2.53×10^{-6} m^2·s^{-1}. The early-time T from the first test (Test A14) after hydraulic fracturing was 8.29×10^{-6} m^2·s^{-1}, which declined to 4.08×10^{-6} m^2·s^{-1} after the first 30 minutes, and continued to decline to 3.16×10^{-6} m^2·s^{-1} during the follow-on test (Test A16). The hydraulic fracturing process probably extended the fracture in Well A into and provided a better hydraulic connection with an area having a higher T value. Hydraulic fracturing records seem to confirm this interpretation in that they suggest that a fracture was propagated early in the hydraulic fracturing process. At late-time a different zone was traversed by that fracture. The decline in T values noted in tests A14 and A16 could be due to dewatering of a shallow aquifer, possibly formed in the reported sandy portion of the surficial material overlying the bedrock in the area of Well B.

Table 9: Comparison of the best pre and post-hydraulic fracturing transmissivity estimates for Wells A and B

Well Name	Pre-Hydraulic Fracturing					Post-Hydraulic Fracturing				
	Test Name	Model Type	Test Period (min)	Transmissivity (m^2·s^{-1})	Source	Test Name	Model Type	Test Period (min)	Transmissivity (m^2·s^{-1})	Source
Well A										
	A2					A14				
		Moench 1 & 3	0 - 80	1.70×10^{-7}	This Study		Moench 1 & 3	0 - 30	8.29×10^{-6}	This Study
		Hantush A/S	60 - 130	8.10×10^{-7}	This Study		Moench 1 & 3	30 - 200	4.08×10^{-6}	This Study
		Hantush A/S	130 - 230	1.78×10^{-6}	This Study		Barenblatt	...	1.21×10^{-5}	[10]

A Reinterpretation of Historic Aquifer Tests of two ...

						Papadopulus-Cooper	...	2.43×10^{-5}	[3]
	A3				A16				
		SVF (F)	20 - 200	5.40×10^{-7}	This Study		Moench 1 & 3	0 - 100	3.16×10^{-6}	This Study
		Moench 1 & 3	200 - 380	1.15×10^{-6}	This Study		Barenblatt	...	1.44×10^{-5}	[10]
						Papadopulus-Cooper	...	1.62×10^{-5}	[3]
	A5				A741				
		Hantush A/S	50 - 500	4.00×10^{-7}	This Study		Moench 1 & 3	0 - 280	1.55×10^{-5}	This Study
		Moench 1 & 3	500 - 800	730×10^{-7}	This Study		Barenblatt	...	2.75×10^{-5}	[10]
		Barenblatt	...	7.20×10^{-7}	[10]		Papadopulus-Cooper	...	6.37×10^{-5}	[3]
		Papadopulus-Cooper	...	2.41×10^{-6}	[3]					
A742									
						Moench 1 & 3	0 - 100	7.00×10^{-6}	This Study
						Barenblatt	...	1.71×10^{-5}	[10]
						Papadopulus-Cooper	...	1.72×10^{-5}	[3]
Well B									
	B3				B25				
		Moench 1 & 3	0 - 80	3.25×10^{-6}	This Study		Moench 1 & 3	0 - 80	6.71×10^{-6}	This Study
		Barenblatt	...	2.87×10^{-6}	[10]		Barenblatt	...	2.20×10^{-6}	[10]
		SHF	...	3.70×10^{-8}	[3]		SHF	...	1.24×10^{-5}	[3]
	B4				B27				
		Moench 1 & 3	0 - 160	6.38×10^{-6}	This Study		Moench 1 & 3	0 - 20	2.37×10^{-6}	This Study
		Barenblatt	...	3.02×10^{-6}	[10]					
		SHF	...	9.30×10^{-8}	[3]					
					B28				
						Moench 3	0 - 16	5.42×10^{-6}	This Study
B743									
						Moench 1 & 3	0 - 55	1.49×10^{-5}	This Study

						Baren-blatt	...	1.49 × 10⁻⁵	[10]
						SHF	...	6.72 × 10⁻⁶	[3]
					B744				
						Moench 1 & 3	0 - 60	6.72 × 10⁻⁶	This Study
						Baren-blatt	...	1.22 × 10⁻⁵	[10]
						SHF	...	1.70 × 10⁻⁵	[3]

Note: Hantush A/S = Leaky aquifer with aquitard storage. SVF (I) = Single vertical fracture (infinite conductivity). SHF = Single horizontal fracture.

A similar pattern was noted during the two 1974 tests of Well A, except that the T values were higher. The T values derived from the Moench model were 1.55×10^{-5} m²·s⁻¹ during the first test (Test A741) and 7.00×10^{-6} m²·s⁻¹ during the second test (Test A742). One explanation for the higher T values is that long-term pumping of the well caused the trough of depression to extend into an area of even higher T values than occurred during the 1973 tests. The subsequent decline in T, again, was probably due to dewatering of the weathered zone. Unfortunately, no information concerning the post-hydraulic fracturing operational history of the well is presented in either the Stewart [10] or dos Santos et al. [3] studies.

Prior to hydraulic fracturing of Well B, the T from the first test (Test B3) was 3.25×10^{-6} m²·s⁻¹. The follow-on test (Test B4) produced a T = 6.38×10^{-6} m²·s⁻¹, an increase that could have been due to better well development, caused by extended pumping of the well (Table 9). There is no evidence of dewatering in the Test B4 data. This is probably due to either the relatively low pumping rates, masking by the effects of well development, or a combination of both factors. After hydraulic fracturing, the calculated T from the first test (Test B25) of 6.71×10^{-6} m²·s⁻¹ is nearly the same as that for Test B4, suggesting no significant change occurred after the hydraulic fracturing procedure. Had dewatering occurred prior to the hydraulic fracturing, this would have indicated that the T value declined immediately after stimulation. The T declines to 2.37×10^{-6} m²·s⁻¹ during the second test (Test B27) and, then, during the final test (Test B28) increases back to 5.42 × 10^{-6} m^2·s^{-1}. In this instance, some dewatering may have occurred during Test B27 followed by expansion of the trough of depression to an area with a higher T during Test B28. The response observed during the first two tests is similar to that which occurred during the two post-hydraulic fracturing tests of Well A.

A similar pattern to the Well A results is noted during the two 1974 tests of Well B. The T obtained using the Moench model is 1.49 × 10^{-5} m^2·s^{-1} during the first test (Test B743) and 6.72 × 10^{-6} m^2·s^{-1} during the second test (Test B744). As with Well A, an explanation for the higher T values is that long-term pumping of the well caused the trough of depression to extend into an area of even higher T and the decline in T during the second test was due to aquifer dewatering. Neither of Stewart [10] or dos Santos et al. [3] provided any information concerning the post-hydraulic fracturing operational history of Well B.

A comparison of the highest T value prior to and the T value immediately after hydraulic fracturing would provide a measure of how effective the process was in each well, by eliminating or minimizing changes due to well development and aquifer dewatering. For both wells, the Moench solutions are used for consistency in making the comparisons. In the case of well A, the highest T value (late-time Test A2) prior to and the T value immediately after (Test A14) hydraulic fracturing were 2.53 × 10^{-6} and 8.29 × 10^{-6}, respectively, or a factor of 3.3, which was similar to the increase in the pumping rates from 38 to 95 L·min^{-1}, or a factor of 2.5. There was no dewatering of fractures and water levels were relatively stable at the end of the tests, suggesting that the pumping rates may provide good approximations of the reliable yields of Well A. For Well B, there was no significant change in the T values just before and immediately after hydraulic fracturing (6.38 × 10^{-6} for Test B4 and 6.71 × 10^{-6} m^2·s^{-1} for Test B25). The water level did not stabilize at 15 L·min^{-1} during Test B4, but was stable at the end of Test B25, while pumping at 37 L·min^{-1}. Primary water-bearing zones were dewatered and water levels never stabilized during each of the follow-on post-hydraulic fracturing tests of

Well B, at pumping rates between 38 and 57 L·min^{-1}. These results indicate that the best estimate is that the yield of Well B increased from less than 15 L·min^{-1} to less than 38 L·min^{-1} as a result of the hydraulic fracturing process. Because there was no increase in the T after hydraulic fracturing in Well B, the increase in yield of Well B may be related to the propagation of the major fracture in Well A hydraulically connecting the two wells after hydraulic fracturing and increasing the drainage area of Well B.

Stewart [10] reported that there was no communication between Wells A and B during pumping tests conducted prior to the hydraulic fracturing procedures. During post-hydraulic fracturing tests, however, there were measurable drawdowns in the non-pumping well that occurred 30 - 60 minutes after the start of pumping. No water level data were included in Stewart [10] so the degree of interference could not be estimated. Because of this evidence that the fracture propagated from Well A and hydraulically connected the two wells, it is likely that the drainage area for Well B was increased, which could account for its increased yield. The sum of the individual yields in the previous paragraph indicate that the total change in yield of the two wells was from less than 53 L·min^{-1} to less than 133 L·min^{-1}, a potential increase by a factor of 2.5, or about one half of the 5.3 times increase indicated by dos Santos et al. [3] . This estimate does not take into account the effects of well interference caused by the hydraulic connection between the two wells. There is insufficient information available to determine the degree of interference, but it is possible that there may have been little or no increase in the total yield of the two wells after the hydraulic stimulations were completed. It would require more extensive testing or long-term monitoring of well operations (records of daily production, hours pumped, and water levels) to determine if this were the case.

SUMMARY AND CONCLUSIONS

The sequential change of the flow regime during the pre-hydraulic fracturing tests of Well A is from early-time wellbore storage (first test, Test A2), through early to mid-time single, vertical, extended fracture (second test, Test A3), to mid to late-time leaky aquifer effects (all tests). These changes and the associated progressively increasing T values are probably due to the effects of well development as a result of the extensive pumping of Well A and the extension of the trough of depression to an area with a higher T. There is an apparent progressive trend of declining T between the individual tests that is probably related to aquifer dewatering. Stewart [10] stated that most of the fractures in the vicinity of the well were steeply dipping. The Moench leaky aquifer model best fits the data from the pumping tests completed immediately after hydraulic fracturing of Well A and the tests conducted one year later. The aquifer test analyses indicate that the one prominent vertical fracture in the well was extended to an area with a higher T value by the hydraulic fracturing process. There are declines in T values during the aquifer tests conducted after hydraulic fracturing in both 1973 and 1974 [10] that are probably related to dewatering of the shallow sandy portion of the aquifer. dos Santos et al. [3] stated that T increased by 46 times after hydraulic fracturing, but the present analyses suggest the T only improved by a factor of 2.8 and that the improvement in T was due to propagation of a fracture to an area of the aquifer with a higher T.

Prior to hydraulic fracturing of Well B, the T nearly doubled between the first and second tests (Tests B3 and B4, respectively), which is probably due to better well development as a result of the extended pumping of the well. There was no evidence of dewatering, possibly due to either the low pumping rates or being masked by the effects of well development. After hydraulic fracturing, there was no immediate change in the T during the first test (Test B25), but it declined and then increased during the second test (Test B27) and third test (Test B28), respectively. One year later, the T again declined after the first test (Test B743) to a value nearly identical with the last pre-hydraulic fracturing test (Test B4), suggesting that there was no effective long-term increase in T due to the hydraulic fracturing process. This would seem to confirm the Stewart [10] observation that the geophysical logs showed no clear evidence that fractures were formed hydraulically in that well. These results differ from those of dos Santos et al. [3] who stated that the T increased in Well B by 285 times after hydraulic fracturing.

The T values after hydraulic fracturing in both Well A and Well B were nearly identical. Because there was no effective change in the T for Well B, it is probable that the primary water-bearing fracture in Well A was propagated in the direction of Well B. Both wells could have then been receiving water from the same reservoir or portion of the aquifer. After hydraulic fracturing, the sum of the individual yields of Well A and Well B increased by a factor of 2.5. However, the wells were then hydraulically connected, so that there may have been no significant increase in the total overall yield of the two wells.

After well development, the two primary flow-controlling mechanisms were leakage and dewatering of fractures. This was not evident on the log-log graphs used by Stewart [10] and the fractures could only be detected on diagnostic and derivative plots. Leakage is identified by a steady drawdown and then recovery of the derivative towards zero. Dewatering of fractures was rateand time-dependent, and produced either a constant drawdown or a peak on a derivative curve. Even with the advantage of an inverse analysis, various models could provide similar fits to specific sets of drawdown data. Application of diagnostic and derivative plots to sort out the best solutions is essential to determining the correct model to apply to a set of time-drawdown data.

ACKNOWLEDGMENTS

We gratefully thank Christoph Butscher of the Karlsruhe Institute of Technology and Ingrid Padilla of the University of Puerto Rico for their careful review of this manuscript. Their suggestions have greatly improved our manuscript.

REFERENCES

1. Theis, C. (1935) The Relation between the Lowering of the Piezometric Surface and the Rate and Duration of Discharge of a Well Using Groundwater Storage. Transactions, the American Geophysical Union, 16, 519-524. http://dx.doi.org/10.1029/TR016i002p00519
2. Cooper, H.H. and Jacob, C.E. (1946) A Generalized Graphical Method for Evaluating Formation Constants and Summarizing Well Field History. Transactions, American Geophysical Union, 27, 526-534. http://dx.doi.org/10.1029/TR027i004p00526
3. dos Santos, J.S., Ballestero, J.S., da Silva, T.P. and Pitombeira, E. (2011) An Analytical Model for Hydraulic Fracturing in Shallow Bedrock Formations. Groundwater, 49, 415-425. http://dx.doi.org/10.1111/j.1745-6584.2010.00727.x
4. Piscopo, V. and Summa, G. (2006) Experiment of Pumping at Constant-Head: An Alternative Possibility to the Sustainable Yield of a Well. Hydrogeology, 15, 679-687. http://dx.doi.org/10.1007/s10040-006-0132-2
5. Batu, V. (1998) Aquifer Hydraulics: A Comprehensive Guide to Hydrogeologic Data Analysis. John Wiley & Sons, Inc., New York.
6. Lebbe, L.C. (1999) Hydraulic Parameter Identification: Generalized Interpretation Method for Single and Multiple Pumping Tests. Springer-Verlag, Berlin.
7. Duffield, G.M. (2007) AQTESOLV for Windows User's Guide. Version 4.5, HydroSOLVE, Inc., Reston.
8. Renard, P., Glenz, D. and Mejias, M. (2009) Understanding Diagnostic Plots for Well-Test Interpretation. Hydrogeology Journal, 17, 589-600. http://dx.doi.org/10.1007/s10040-008-0392-0
9. Beauheim, R.L., Roberts, R.M. and Avis, J.D. (2004) Well Testing in Fractured Media: Flow Dimensions and Diagnostic

Plots. Journal of Hydraulic Research, 42, 69-76. http://dx.doi.org/10.1080/00221680409500049

10. Stewart, G.W. (1974) Hydraulic Fracturing of Drilled Water Wells in Crystalline Rocks of New Hampshire. Tech. Rep., New Hampshire Department of Resources and Economic Development, Concord.

11. Driscoll, F.G. (1986) Groundwater and Wells. 2nd Edition, Johnson Division, St. Paul.

12. Boulton, N.S. (1954) The Drawdown of the Water Table under Non-Steady Conditions near a Pumped Well in an Unconfined Formation. Proceedings of the Institute of Civil Engineers, 3, 564-579.

13. Boulton, N.S. (1963) Analysis of Data from Non-Equilibrium Pumping Tests Allowing for Delayed Yield from Storage. Proceedings of the Institute of Civil Engineers, 26, 469-482. http://dx.doi.org/10.1680/iicep.1963.10409

14. Boulton, N.S. (1973) The Influence of Delayed Drainage on Data from Pumping Tests in Unconfined Aquifers. Journal of Hydrology, 19, 157-169.

15. Neuman, S.P. (1972) Theory of Flow in Unconfined Aquifers Considering Delayed Response of the Water Table. Water Resources Research, 8, 1031-1045. http://dx.doi.org/10.1029/WR008i004p01031

16. Neuman, S.P. (1974) Effect of Partial Penetration on Flow in Unconfined Aquifers Considering Delayed Gravity Response. Water Resources Research, 10, 303-312. http://dx.doi.org/10.1029/WR010i002p00303

17. Neuman, S.P. (1975) Analysis of Pumping Test Data from Anisotropic Unconfined Aquifers Considering Delayed Gravity Response. Water Resources Research, 11, 329-342. http://dx.doi.org/10.1029/WR011i002p00329

18. Streltsova, T.D. (1975) Hydrodynamics of Groundwater Flow in a Fractured Formation. Water Resources Research, 12, 405-414. http://dx.doi.org/10.1029/WR012i003p00405

19. Boulton, N.S. and Streltsova, T.D. (1977) Flow to a Well in an Unconfined Fractured Aquifer. In: Dilamarter, R.R. and Csallany, S.C., Eds., Hydrologic Problems in Karst Regions, University of Western Kentucky, Bowling Green, 214- 227.
20. Boulton, N.S. and Streltsova, T.D. (1977) Unsteady Flow to a Pumped Well in a Two-Layered Water-Bearing Formation. Journal of Hydrology, 35, 245-256. http://dx.doi.org/10.1016/0022-1694(77)90004-X
21. Boulton, N.S. and Streltsova, T.D. (1977) Unsteady Flow to a Pumped Well in a Fissured Water-Bearing Formation. Journal of Hydrology, 35, 257-270. http://dx.doi.org/10.1016/0022-1694(77)90005-1
22. Boulton, N.S. and Streltsova, T.D. (1978) Unsteady Flow to a Pumped Well in a Fissured Aquifer with a Free Surface Level Maintained Constant. Water Resources Research, 14, 527-532. http://dx.doi.org/10.1029/WR014i003p00527
23. Moench, A.F. (1984) Double Porosity Models for a Fissured Groundwater Reservoir with Fractured Skin. Water Resources Research, 20, 831-846.
24. Gringarten, A.C., Ramey Jr., H.J. and Ragavan, R. (1975) Applied Pressure Analysis for Fractured Wells. Journal of Petroleum Technology, 27, 887-892.
25. Barker, J.A. (1988) A Generalized Radial Flow Model for Hydraulic Tests in Fractured Rock. Water Resources Research, 24, 1796-1804. http://dx.doi.org/10.1029/WR024i010p01796
26. Hantush, M.S. (1960) Modification of the Theory of Leaky Aquifers. Journal of Geophysical Research, 65, 3713-3725. http://dx.doi.org/10.1029/JZ065i011p03713
27. Moench, A.F. (1985) Transient Flow to a Large-Diameter Well in an Aquifer with Storative Semiconfining Layers. Water Resources Research, 21, 1121-1131.
28. en, Z. (1995) Applied Hydrogeology for Scientists and Engineer. CRC Press, Boca Raton.

29. Samani, N., Pasandi, M. and Barry, D.A. (2006) Characterizing a Heterogeneous Aquifer by Derivative Analysis of Pumping and Recovery Test Data. Journal of Geological Society of Iran, 1, 29-41. http://infoscience.epfl.ch/record/118466/files/jp142.pdf
30. Bourdet, D., Ayoub, J.A. and Pirard, Y.M. (1989) Use of Pressure Derivative in Well-Test Interpretations. SPE Formation Evaluation, 4, 293-302.http://dx.doi.org/10.2118/12777-PA
31. van Tonder, G.J., Botha, J.F., Wen-Hsing, C., Kunstmann, H. and Xu, Y.X. (2001) Estimation of the Sustainable Yields of Boreholes in Fractured Rock Formations. Journal of Hydrology, 241, 70-90. http://dx.doi.org/10.1016/S0022-1694(00)00369-3
32. Sauveplane, C. (1984) Pumping Test Analysis in Fractured Aquifer Formations: State of the Art and some Perspectives. In: Rosenshein, J. and Bennett, G.D., Eds., Groundwater Hydraulics. Water Resources Monograph 9, American Geophysical Union, Washington DC, 171-206.
33. Johns, R.A., Semprini, L. and Roberts, P.V. (1992) Estimating Aquifer properties by Nonlinear Least-Squares Analysis of Pump Test Response. Groundwater, 30, 68-77.http://dx.doi.org/10.1111/j.1745-6584.1992.tb00813.x
34. Carrera, J. and Neuman, S.P. (1986) Estimation of Aquifer Parameters under Transient and Steady State Conditions: 2. Uniqueness, Stability, and Solution Algorithms. Water Resources Research, 22, 211-227. http://dx.doi.org/10.1029/WR022i002p00211
35. Sun, N.Z. (1994) Inverse Problems in Groundwater Modelling. Kluwer Academic Publishers, Dordrecht.
36. Madsen, K., Nielsen, H.B. and Tingleff, O. (2004) Methods for Non-Linear Least Squares Problems. 2nd Edition, Informatics and Mathematical Modelling (IMM), Technical University of Denmark (DTU), Lyngby.http://www2.imm.dtu.dk/pubdb/views/edoc_download.php/3215/pdf/imm3215.pdf

37. Haupt, R.L. and Haupt, S.E. (2004) Practical Genetic Algorithms. 2nd Edition, John Wiley & Sons, Inc., Hoboken.
38. Saleem, Z.A. (1970) A Computer Method for Pumping-Test Analysis. Groundwater, 8, 21-24. http://dx.doi.org/10.1111/j.1745-6584.1970.tb01318.x
39. Jacquard, P. and Jain, C. (1965) Permeability Distribution from Field Pressure Data. Society of Petroleum Engineers Journal, 5, 281-294. http://dx.doi.org/10.2118/1307-PA
40. Jahns, H.O. (1966) A Rapid Method for Obtaining a Two-Dimensional Reservoir Description from Well Pressure Response Data. Society of Petroleum Engineers Journal, 6, 315-327. http://dx.doi.org/10.2118/1473-PA
41. Sayed, S.A.S. (1990) Automated Analysis of Pumping Tests in Unconfined Aquifers of Semi-Infinite Thickness. Groundwater, 28, 108-112. http://dx.doi.org/10.1111/j.1745-6584.1990.tb02234.x
42. Boonstra, H. and Soppe, R. (2006) Well Hydraulics and Aquifer Tests. In: Delleur, J.W., Ed., The Handbook of Groundwater Engineering, 2nd Edition, CRC Press, Taylor and Francis Group, Boca Raton, 10-1-10-35.
43. Williamson, W.H. and Woolley, D.R. (1980) Hydraulic Fracturing to Improve the Yields of Bores in Fractured Rock. Tech. Rep. 55, Australian Water Resources Council, Australian Government Publishing Service, Canberra.
44. Herbert, R., Talbot, J.C. and Buckley, D.K. (1993) A Study of Hydraulic Fracturing Used on Low Yielding Boreholes in the Crystalline Basement Rocks of Masvingo Province, Zimbabwe. In: Memoires of the 24th Congress, International Association of Hydrogeologists, 28th June-2nd July 1993, Ås (Oslo), 698-726.
45. Gale, J.E. and MacLeod, R. (1995) Assessing the Effectiveness of Fracture Stimulation for Increasing Well Yield in Newfoundland. Tech. Rep. Canada-Newfoundland Agreement Respecting Water Resource Management, Government of Newfoundland and Labrador, Department

of Environment, Water Resources Division, Environment Canada, Environmental Conservation Strategies Division.

46. Montgomery, C.T. and Smith, M.B. (2010) Hydraulic Fracturing: History of an Enduring Technology. Journal of Petroleum Technology, 62, 26-32.http://www.spe.org/jpt/print/archives/2010/12/10Hydraulic.pdf

47. Benzie, F., Chenier, F., Kleiman, H., McDonald, H., Surrell, J., Thomas, C. and Van Alstine, J. (1996) Hydraulic Fracturing Request Review Policy. Tech. Rep. Policy Review, Michigan Department of Environmental Quality, Lansing.http://www.michigan.gov/documents/deq/deq-wd-gws-wcu-hydraulicfracturing_270750_7.pdf

48. Gringarten, A.C. and Ramey, H.J. (1974) Unsteady State Pressure Distributions Created by a Well with a Single Horizontal Fracture, Partial Penetration or Restricted Entry. Journal of the Society of Petroleum Engineers, 14, 413-426.

49. Murdoch, L.C. and Slack, W.W. (2002) Forms of Hydraulic Fractures in Shallow Fine-Grained Formations. Journal of Geotechnical and Geoenvironmental Engineering, 128, 479-487. http://dx.doi.org/10.1061/(ASCE)1090-0241(2002)128:6(479)

Chapter 5

Evaluation of Methane Yield on Mesophilic-Dry Anaerobic Digestion of Piggery Manure Mixed with Chaff for Agricultural Area

Dong-Heui Kwak[1], Mi-Sug Kim[1], Jae-Seung Kim[2], Young-Youl Oh[3], Soon-Ok Noh[4,5], Byung-Ok So[4,5], Su-Young Jung[4,5], Su-Jin Jung[4,5], and Soo-Wan Chae[4,5]

[1]Department of Environmental and Chemical Engineering, Seonam University, Namwon, Republic of Korea

[2]Department of Environmental Engineering, Chonbuk National University, Jeonju, Republic of Korea

[3]Yoyo Korea Agricultural Association, Jeongehb, Republic of Korea

[4]Clinical Trial Center for Functional Foods, Chonbuk National University Hospital, Jeonju, Republic of Korea

[5]Department of Medical Nutrition Therapy, Chonbuk National University, Medical School, Jeonju, Republic of Korea

ABSTRACT

A mesophilic-dry anaerobic digestion process is valid in treating high-concentration substrates containing low moisture content. It has merits of lower wastewater discharge and lower heat capacity required in maintaining reactor temperature as compared with a thermophilic-wet anaerobic digestion process. In fact, chaff can be easily obtained in farming areas and used as a mixture substrate as one of bulking agents for controlling moisture and supplying carbon. For this reason, this study applies the chaff to improve livestock manure, which contains high moisture content and is discharged from domestic pig farms. This study aims at verifying its feasibility for improving methane production efficiency on a basis of BMP (Biochemical Methane Potential) assay obtained through a series of experiments. Finding results were methane gas production and gas production per volatile solid (VS) added, and methane gas production among biogas production were increased as the chaff added in the piggery manure was increased. According to experimental results for improving the methane production efficiency, mixture of the chaff and the piggery manure played an important role in controlling the moisture content and improving the methane gas production rate, and also verified its feasibility in the mesophilic-dry anaerobic digestion process indicating relatively less difficulty for operation and management.

INTRODUCTION

A methane fermentation process has an advantage in treating organic contaminants for preventing environmental pollution when comparing a conventional aerobic treatment process. Naturally, the methane fermentation process has a combination limit of processes but it is relatively useful in aspects of energy production and resource collection. The methane fermentation process such as an anaerobic digestion process is a skill studied and used for a long time and is recently being magnified in a situation as an international concern focusing on climate change control and renewable energy demand. Especially, a biogas plant, one of the methane fermentation skills, has been used in many countries and known as one of effective strategy techniques for bio-fuel production [1].

Domestic livestock manure emission classified by livestock types has a component ratio as follows; 57.6% piggery manure (740,000 m^3/d) and 42.4% cow manure (540,000 m^3/d) [2], and dairy cow manure of the cow manure is emitted in the overcrowded area such as farms but few Korean native cattle are raised in small farmers and there are many bad cases in collecting and in treating Korean native cattle manure. The organic content is a raw matter to produce the methane and exists in the piggery manure as low as 2% to 5%. Thus, utilization of the piggery manure as a substrate is low and also it is known well that a stable operation of an anaerobic digestion tank is difficult because a fluctuation range of the organic content is large periodically and the moisture content is high enough [3].

The anaerobic digestion process is divided into a wet process and a dry process according to solid content or moisture content of the substrate used. Until the mid- 1980s, the wet process has been mainly applied in the field using waste matters within 10% solid content as the substrate. With EU as the center rapidly from the 1990s, however, the dry process has been developed to digest organic waste matters containing the solid content over 20% [4]. In treating the high-concentration substrate having low moisture content, the dry process requires low heat capacity to maintain the reactor temperature and discharges low wastewater after treatment [5]. However, it is not valid to put the livestock manure into the dry process directly because the livestock manure emitted from domestic piggery farms contains a great deal of moisture and the solid content of the livestock manure is very low. Meanwhile, the bulking agents such as rice straws, chaff, dead leaves fragments, sawdust, etc. are easy to obtain in the farm area and such agricultural byproducts have been used as the bulking agent for composing manure from old times and also as the carbon supplement for maintaining the proper C/N ratio. Practically, most of domestic livestock farms are located in the farming settlement that is producing a great deal of the bulking agent. In adopting the biogas plant in the domestic farming areas, the dry anaerobic digestion process is in a more advantageous situation than the wet process when considering realistic conditions, In Europe recently, studies on the dry digestion operation for the municipal organic solid waste are actively proceeding to reduce waste amounts for landfill and to produce the bio-energy [6,7]. However, previous studies mainly present that the operation results for high temperature (50°C - 60°C) conditions and continuous operation cases are also pretty rare [8].

To analyze the ultimate methane production rate (mL/gVS$_{added}$) caused by organic matters as the substrate for the anaerobic digestion process, it measures the methane amounts produced during the anaerobic batch incubation period and cumulative methane formula can be used to determine the methane production yield based on the observed data. Representative models such as Modified Gompertz model or Exponential model are used to analyze experiment data obtained through the methane production potential test [9,10]. Using those models described in Equations (1) and (2) as below, comparative studies are variously proceeding to determine the ultimate methane production yield of the substrates related to diverse components [11,12].

Modified Gompertz Model Equation [13]:

$$M = M_o \times \exp\left\{-\exp\left[\frac{R_m \times e}{M_o}(\lambda - t) + 1\right]\right\} \quad (1)$$

where, M: cumulative methane production yield (mLCH$_4$/g-VS)

t: incubation time of an anaerobic digestion tank (days)
M_o: ultimate methane production yield (mL-CH_4/g-VS)
R_m: maximum methane production rate (mL-CH_4/gVS·day)
e: exp (1) = 2.71828182
λ: lag phase, days Exponential Model Equation:

$$B = B_o\left(1-e^{-kt}\right) \quad (2)$$

where, B: cumulative methane production yield (mLCH_4/g-VS)
t: incubation time of an anaerobic digestion tank (days)
B_o: ultimate methane production yield (mL-CH_4/g-VS)
k: 1st order reaction rate constant (day^{-1}).

With the purpose of energy resource recovery through the methane production due to the piggery manure in the farming area, this study conducts a series of experiments with the dry anaerobic digestion using the chaff. The chaff is easily obtained in the farming area as one of the byproducts and as the substrate to mix with the manure as well as to control the moisture. The anaerobic digestion has been conducted in the single-phase mesophilic condition, BMP (biochemical methane potential) assay has been applied to estimate the methane production potential due to the several mixture ratios between the piggery manure and the chaff based on the experimental results, and this study has been accomplished to find a way to improve the methane production efficiency.

MATERIALS AND METHODS

Experimental Equipments and Operation Conditions

In experiments, a batchwise reactor of methane yield is prepared for single-phase anaerobic digestion as shown in Figure 1. For a dry methane production process of livestock manure as a main substrate, typical experimental conditions were adopted to examine gas production yield and responses characteristics. Experiments were set up to control pH if necessary and to incubate for 40 days as controlling to keep typical temperature for mesophilic digestion in a range of 35°C ± 1°C [14], and an additional agitator was excluded in the experiments.

Figure 1 describes a schematic diagram of a batchwise single phase digester for methane yield. 0.5 L serum bottles were set up to an incubator at a constant temperature. Operation conditions were monitored for every serum bottle in different mix proportions between piggery manure as a main substrate and chaff as a mixture substrate to control the moisture. Major items such as pH change and the gas production yield from the serum bottle were monitored due to each condition. The operator also measured generating-capacity and methane content in gas collected in a teflon bag through an exhaust pipe in the middle of a gas-tight rubber stopper of the serum bottle at a constant time interval, every 3 days. Additionally, taking small amounts from all of samples and putting them into each 50 mL-tube, the operator measured its weight per VS (volatile solid) changed with operating under the same conditions.

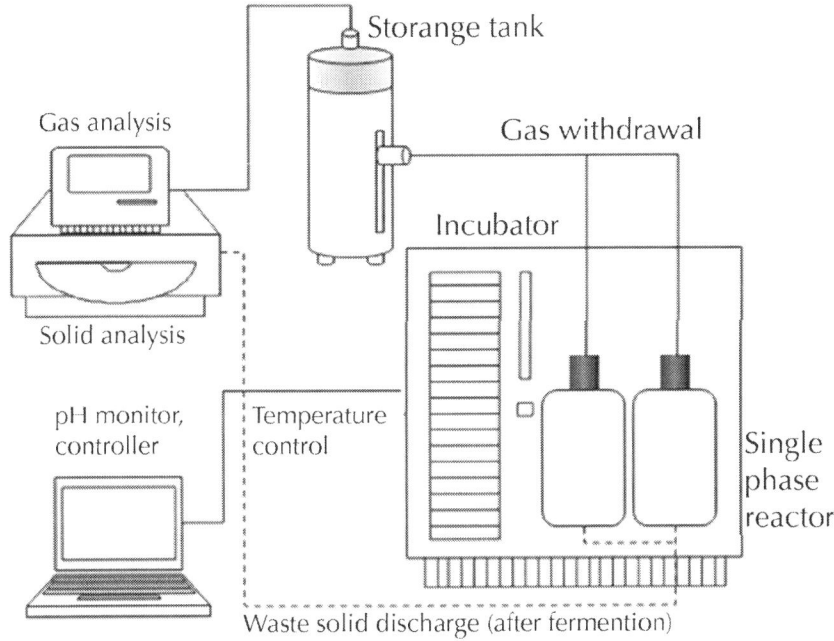

Figure 1: Schematic diagram of batchwise single phase digester for methane yield.

Table 1 presents operation conditions and a substrate composition in the single phase digester. For two operation parameters such as the substrate concentration and the solid content, the methane production was estimated in different mixing ratios between the piggery manure and the chaff. All samples except a control group (marked as Run 1) were added trace elements in order to minimize unstable effects of microbial growth due to lack of essential elements in the anaerobic digester.

Sample and Analysis Method

Livestock manure has been concerned in domestic water management as a non-point source. Among the livestock manure, piggery manure is newly applied as a main substrate for methane production. To improve the methane production, chaff in powder form (below than 100 mesh) is also mixed with the piggery manure to control the moisture as well as to improve the methane production. Major properties of the piggery manure are shown in Table 2 and chemical properties of all samples (Run 1 - Run 5) are described in Table 3. Also, Table 4 presents trace elements (mineral salts and trace metals) added for the safe operating condition. Water quality and solid matters were analyzed by the standard method (2005) in this study.

The piggery manure samples were collected in the retaining tank for gathering and mixing the manure of the pigsty before inputting washing water of a pig farm.

Table 1: Operation condition and substrate composition of single phase digester

Samples	Mixing ratio between manure and chaff	Substrate and solid content			Trace elements
		Piggery manure (g)	Powder of chaff* (g/L)	Solid content* (%)	
Run 1	2.50:1 (control)	300	120	35.4	non-spiked
Run 2	4.17:1	300	72	27.2	spiked
Run 3	3.13:1	300	95	30.6	spiked
Run 4	2.78:1	300	108	33.1	spiked
Run 5	2.50:1	300	120	35.7	spiked

Note: *Converted value into concentration.

Table 2: Chemical composition of piggery wastewater used as main substrate in this study

Description	Unit	Measured values
pH		8.7

BOD	mg/L	2205
CODcr	mg/L	2221
SCODcr	mg/L	1126
T-N	mg/L	3439
T-P	mg/L	121.8
NH3-N	mg/L	1304
P_4^{3-}-P	mg/L	4.3
Alkalinity	mg/L as CaCO3	5800
Fixed solid	mg/L (%)	5900 (0.59)
Volatile solid	mg/L (%)	3300 (0.33)

Table 3: Chemical composition of experimental substrates (mixture of piggery manure and chaff)

Description	Unit	Run 1	Run 2	Run 3	Run 4	Run 5
pH		8.1	7.9	7.8	8.0	8.0	8.0
COD_{cr}	mg/kg	16,587	10,092	14,309	15,447	16,760
T-N	mg/kg	1583	3286	2988	1848	1643
T-P	mg/kg	151.4	189.2	169.2	159.2	153.2
NH_3-N	mg/kg	1122.6	1553.1	1416.0	1401.2	1204.8
Fixed solid	%	4.071	6.129	5.707	5.147	4.571
Volatile solid	%	3.048	2.984	2.879	2.966	3.071

Characteristics of target wastewater compiled from the piggery farm were indicated high pH around 8.7 and high alkalinity about 5800 mg/L as $CaCO_3$, which were typical in the livestock wastewater. The wastewater was measured in BOD (2205 mg/L) and $TCOD_{Cr}$ (2221 mg/L) and contained soluble COD_{Cr} ratio around 50.7%. Total solid concentration was 0.92% with 99.1% water content and the solid consisted of VS (35.9%) and inorganic solid (64.1%). Fraction of nutrients compared with the organic was as follows; $COD:NH_3\text{-}N:PO_4^{3-}\text{-}P = 51.7:30.3:1.0$.

Table 3 describes chemical compositions of experimental substrates in five different sample groups from Run 1 to Run 5 classified by different mixing ratios between the piggery manure and the chaff at identical conditions. The solid content was ranged from 27.2% to 35.4% and COD_{cr} was distributed from 10.092 mg/L to 16,760 mg/L. Ratios of the organic matter and nutrients (COD:T-N:T-P) for the sample groups (Run 1 - Run 5) were 109.6:10.5:1, 53.3:17.4:1, 84.6: 17.7:1, 97.0:11.6:1, and 109.4:10.7:1, respectively.

BMP Assay

A BMP (biochemical methane potential) assay was developed by Owen et al. [10] to evaluate potential efficiency for biodegradability of target livestock manure in an anaerobic process and it was analyzing organic concentration converted into CH_4. In this study, a serum bottle was filled with a target sample, covered with a butyl rubber septum, sealed with a reinforced plastic lid, and kept in an incubator at a constant temperature 35°C to induce anaerobic degradation. Gas producing capacity and its composition were analyzed every time interval and then a methane production rate was recorded. The control group (Run 1) was examined under the same conditions to modify other gas capacity and other effects occurred from the target experimental samples. Unlike the target experimental groups, the control group did not add trace elements consisting of mineral salts and trace metals. The gas producing capacity was measured using a glass syringe in the constant time interval and the methane production rate was analyzed using Gas Chromatography-Mass Spectrometer, GC-MS (6890 N Network GC system).

Table 4: Composition of mineral salts and trace metals

Concentration of mineral salts (mg/L)				Concentration of trace metals (mg/L)	
NH_4Cl	0.53	$MnCl_2 \cdot 4H_2O$	0.0005	$NaMoO_4 \cdot 2H_2O$	0.00001
$CaCl_2 \cdot 2H_2O$	0.075	H_3BO_3	0.00005	$CoCl_2 \cdot 6H_2O$	0.0005
$MgCl_2 \cdot 6H_2O$	0.1	$ZnCl_2$	0.00005	$NiCl_2 \cdot 6H_2O$	0.00005
$FeCl_2 \cdot 4H_2O$	0.02	$CuCl_2$	0.00003	Na_2SeO_3	0.00005

RESULT AND DISCUSSION

Experimental Result of Dry Mix Digestion

Piggery manure has had a difficult time in proceeding wet anaerobic digestion because of its high moisture content. For improving the anaerobic digestion of the piggery manure, it may be mixed with chaff obtained easily in farming areas. Table 5" target="_self"> Table 5 presents main experimental results of gas production and methane yields through 40 days digestion operating period for five sample groups.

In anaerobic degradation reactions of organic matters, substrate concentrations and physical and chemical compositions can affect on a reaction rate of hydrolysis and acid formation. Hence, this experiment set an equal amount of the livestock manure for all of samples to prevent the methane production rate from reducing.

Table 5: Experiment results of methane yields for five types of mixed substrates

Descriptions	Samples				
	Run 1	Run 2	Run 3	Run 4	Run 5
Cumulative volume of biogas yield (L)	3.21	3.48	3.69	4.19	4.76
Average daily biogas yield (L)	0.708	0.780	0.825	0.933	1.040

Maximum daily biogas yield (L)	1.070	1.160	1.230	1.397	1.587
Average methane fraction in biogas (%)	0.639	0.680	0.692	0.689	0.720
Maximum methane fraction in biogas (%)	0.735	0.782	0.791	0.842	0.849
Volatile solid (added) (g)	12.8	11.1	11.4	12.1	12.9
VS removal fraction (-)	0.667	0.843	0.808	0.833	0.797
Total biogas production rate (L/g-VSremoved)	0.376	0.372	0.396	0.416	0.463
Average daily biogas production rate (L/g-VSremoved·day)	0.102	0.109	0.121	0.116	0.139
Maximum daily biogas production rate (L/g-VSremoved·day)	0.198	0.201	0.208	0.216	0.235
Terminate duration of methane yield (days)	37 - 40	34 - 37	37 - 40	37 - 40	37 - 40

For considering the biogas production as shown in Table 5, total gas production and the gas production yield per VS added were also increased when the more chaff was mixed with the piggery manure. As mentioned above, the methane gas fraction in the biogas swung upward as the more chaff was added in the mixture. The addition of the chaff obviously improved the methane production rate and reduced the moisture content. In Figure 5, Run 1 and Run 2 used the constant mix proportion of the chaff but Run 1 without the trace elements produced less gas amounts compared to Run 2.

VS removal efficiency was estimated depending on measurement results of VS change amounts and was distinct from the change of the gas production yield. Although the VS was measured for the same sample under the identical conditions, in fact, there was a limitation in conforming tiny change of weight when the 0.5 L-serum bottle was measured directly. Thus, it may be possible to have some error because this study has measured a 0.05 L-additional container at the constant time interval.

Figure 2 indicates cumulative gas production rates of five sample groups for the digestion period. A large amount of gas was produced when the large amount of the chaff was added. Also, the daily gas production rate was slightly rapid at the sample group with the large amount of the chaff as shown in Figure 3 and a maximum gas production peak was appeared at 10 - 15 days after the digestion was started. Therefore, the sample group with the larger chaff content indicated the more gas production yield and the faster gas production rate, and the earlier maximum gas production peak.

Ultimate Methane Productivity Evaluation

Theoretical total methane productivities of piggery manure (516 L/g-VS by the hog and 530 L/g-VS by the sow) were higher than those of the cow manure (469 L/g-VS). The ultimate methane yield was in order as follows; 356 L/g-VS by the hog, 275 L/g-VS by the sow, and 148 L/g-VS by the milk cow. Also, the straw and the chaff were known to have much more methane yields as compared with the animal manure [15].

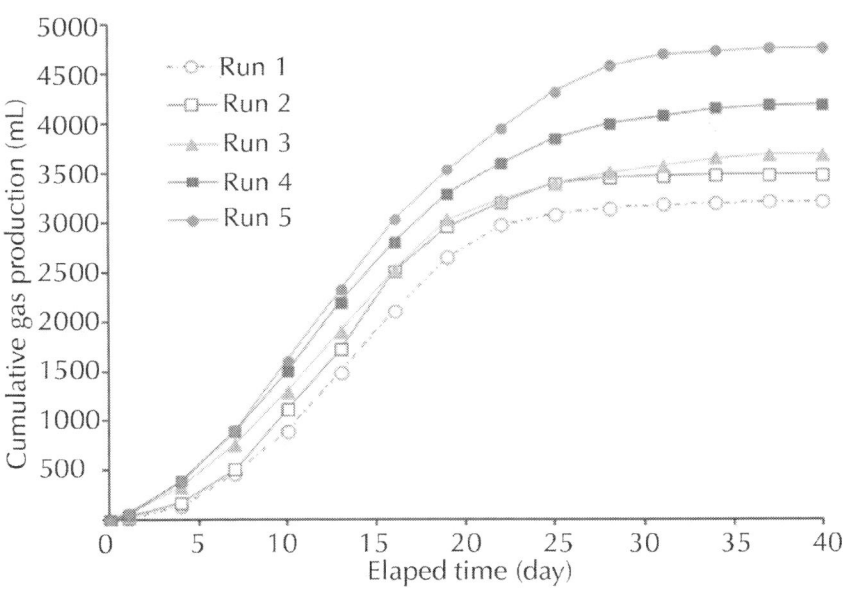

Figure 2: Comparison of cumulative gas yield from five types of samples.

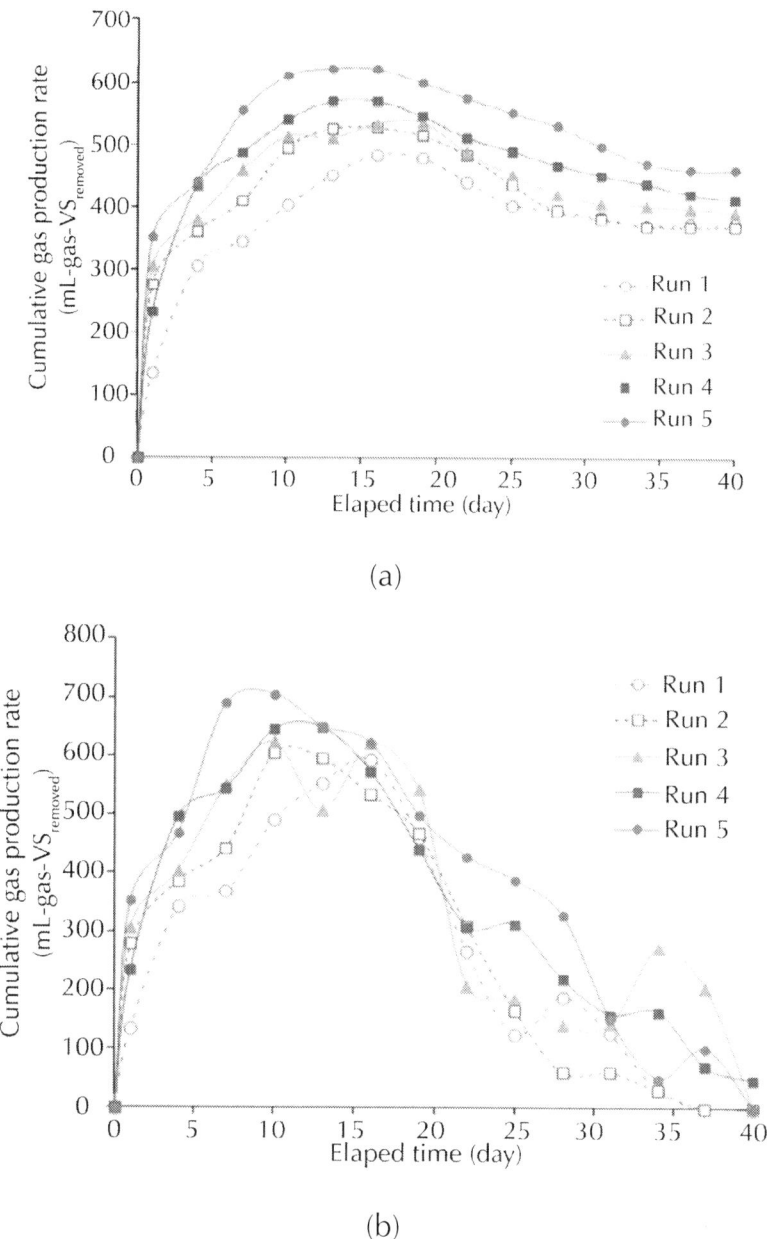

Figure 3: Distribution of gas production rate from five types of samples (3 days period). (a) Cumulative gas production rate; (b) Daily gas production rate.

Modified Gompertz Model (MGM) and Exponential Model (EM) were used to evaluate the cumulative methane yield on a basis of the BMP assay operated for 40 days under the experimental conditions for five sample groups. By comparison of the ultimate methane yields (cumulative methane yields at the final time, 40 days) evaluated by two models, MGM (168.39 - 314.08 mLCH$_4$/g-VS$_{added}$) determined relatively lower ranges than EM (167.62 - 312.47 mL-CH$_4$/g-VS$_{added}$).

In Table 6, for the mesophilic-dry digestion experiment, the ultimate methane yield was observed in the range of 163.02 mL - 313.27 mL-CH$_4$/g-VS$_{added}$ which was applied in two models as M_o and B_o. The methane yield was also increased when the chaff mixing ratio was high as shown in the case of total gas production rate. Also, Run 1 without the trace elements indicated the lower cumulative methane yield (77.15 mL-CH$_4$/g-VS$_{added}$) than that for Run 2 with the trace elements.

Table 6: Summary of kinetic parameters predicted by two models

Model	Parameters			Samples		
		Run 1	Run 2	Run 3	Run 4	Run 5
	Cumulative methane yield, M (mL-CH4/g-VS,dd$_e$d)	168.49	245.33	256.65	291.22	314.08
	Ultimate methane yield, $M_.$ (mL-CH4/g-VS,dd$_e$d)	168.02	245.17	256.03	291.57	313.27
Modified Gompertz Model						
	Maximum methane production rate, $R_.$, (mL-CH4/g-VSthi$_e$d•day)	15.98	16.97	16.90	17.58	19.77
	Lag phase, 1 (day)	3.0	3.0	3.0	3.0	3.0
	Determination coefficient, R^2 (-)	0.94	0.98	0.99	0.99	0.98

Exponential Model	Cumulative methane yield, B (mL-CH_4/g-$VS_a dd_e d$)	167.62	245.04	254.79	291.14	312.47
	Ultimate methane yield, B. (mL-CH_4/g-$VS_a dd_e d$)	168.02	245.17	256.03	291.57	313.27
	Reaction rate constant, k (May)	0.15	0.19	0.13	0.16	0.15
	Determination coefficient, R^2 (-)	0.89	0.84	0.92	0.85	0.871

Note) $^*M_o$ or B_o is the ultimate methane yield observed during the experiment.

The results of BMP assay simulated by two types of models showed a difference and the modified Gompertz Model was better fit compared with the Exponential Model as shown in Figure 4.

In the experiment, the ultimate methane yield per the inserted VS was distributed in the range of 163.02 mL - 313.27 mL-CH_4/g-VS_{added}. In previous studies, mixture digestion of food waste and livestock manure was tested in the thermophilic-wet digestion process and the ultimate methane yield per the inserted VS was ranged from 313.35 to 377.43 mL-CH_4/g-VS_{added} [16], which was relatively higher than that obtained in this study, and similar to that of the single livestock manure tested at the thermophilic digestion and its result presented as 241 mL/g-VS_{added} [17]. In addition, typical experimental values obtained from foreign countries were 250 L/g-VS for the cow manure including the straw and 279 L/g-VS [18] for the horse-manure with the straw estimated by the thermophilic-wet anaerobic digestion process, and 318 mL/g-VS_{added} [19] for sewage sludge in similar. However, comparative evaluation represented little because BMP assay were varied greatly due to the operation conditions such as nutriment component and addition amount, heat, agitation, and so on [20].

According to the evaluation result of the ultimate methane yield, the mixture of the livestock manure and the chaff in the mesophilic anaerobic digestion process has an advantage when comparing with the only livestock manure containing high moisture content at the thermophilic digestion process in common. Also, it is verified that the trace elements are required to add when the chaff to mix with the livestock manure is not sufficient. Therefore, further study is required more detailed research.

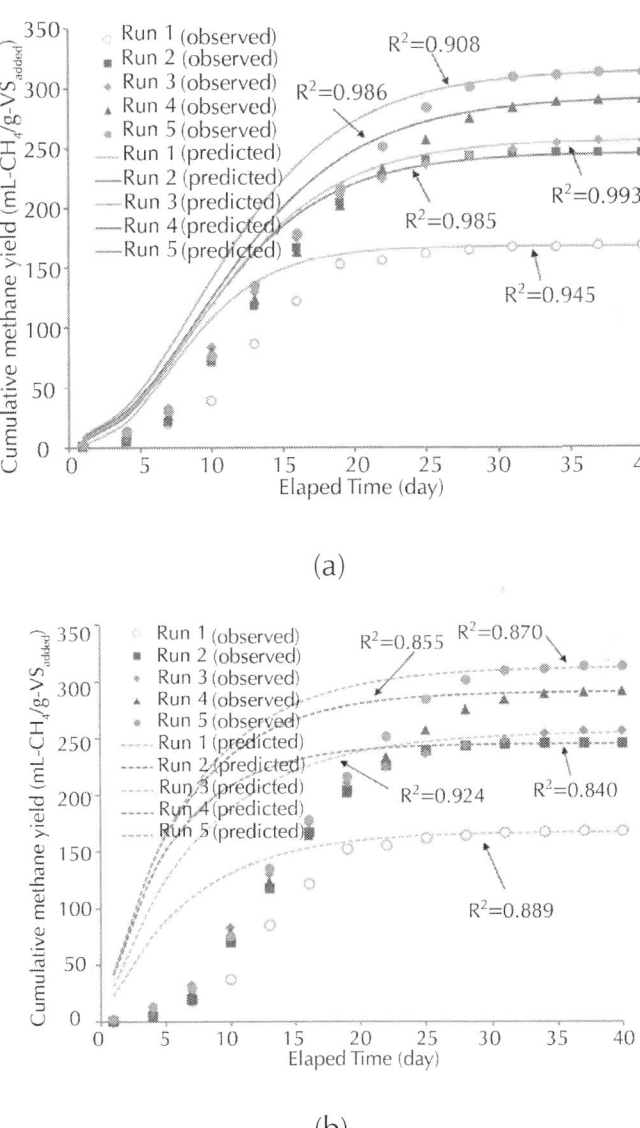

(a)

(b)

Figure 4: Biochemical methane potential assay simulated by two models. a) Modified Gompertz Model; b) Exponential Model.

Mixture Ratio of Manure and Chaff

On the domestic side, food waste as a mixture substrate of livestock manure has been mainly used to solve a pending issue and to improve methane production efficiency. Domestic studies related to optimal mixing ratio of the livestock manure and the food waste have been conducted but the study using herbal plants such as chaff and straw is rare. Because rural areas mainly produce the livestock manure and urban areas discharge the food waste mostly, facilities and energy for transportation are required if the food waste is considered as the substrate mixing with the animal manure. The herbal plants like the chaff are easily obtained in the rural area and mostly used to mix with the livestock manure as shown in developed countries having many cases for biogas plants [21].

With that background, this study verified that the methane production rate was varied due to the amount of the chaff mixed with the livestock manure. As the amount of the chaff was enlarged to expand biogas production and to increase methane production efficiency, it was naturally converted to the dry anaerobic digestion process caused by the lower moisture content and it had little regard for wastewater emission. Coverse et al. studied on the methane production using the high rate of organic loading as inserting carbohydrate mixture compounds [22]. According to their study, the maximum methane production rate was high at the thermophilic digestion but overall methane production yield was larger at the mesophilic digestion. A previous study applied the straw as the mixture substance had reported the most methane production was yielded from the mixture sample of 3% straw and the livestock manure containing 5% solid [23].

To verify the optimal mixing ratio of the piggery manure (SM) as the main compound and the chaff (SB) as the additional mixing compound in this study, the ultimate methane production yield and the maximum production rate due to SM/SB ratio are presented in Figure 5. As shown in Figure 5(a), the ultimate methane production yield and the maximum methane production rate were increased less than 3.5 of SM/SB ratio. The SM/SB ratio is remarkably varied due to the livestock manure composition changed by the stock farmers and the discharging season (or time) and its role may be slightly changed due to the type of the chaff and the condition of the chaff. Therefore, it is necessary to develop a standard model using the more accurate SM/SB ratio examined by several parameters such as solid content, livestock ma nure types, chaff types, dry conditions of the chaff including moisture content and woody presence, and so on.

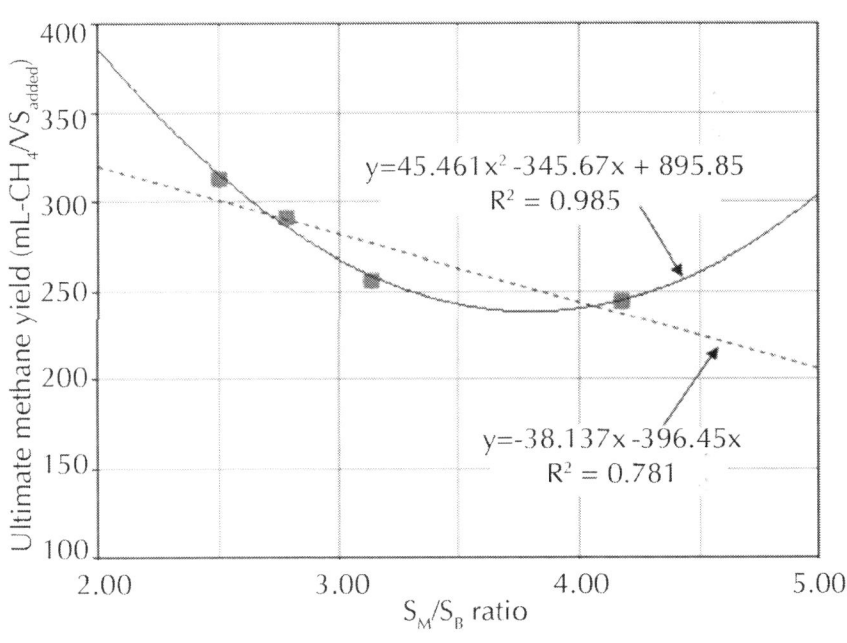

(a)

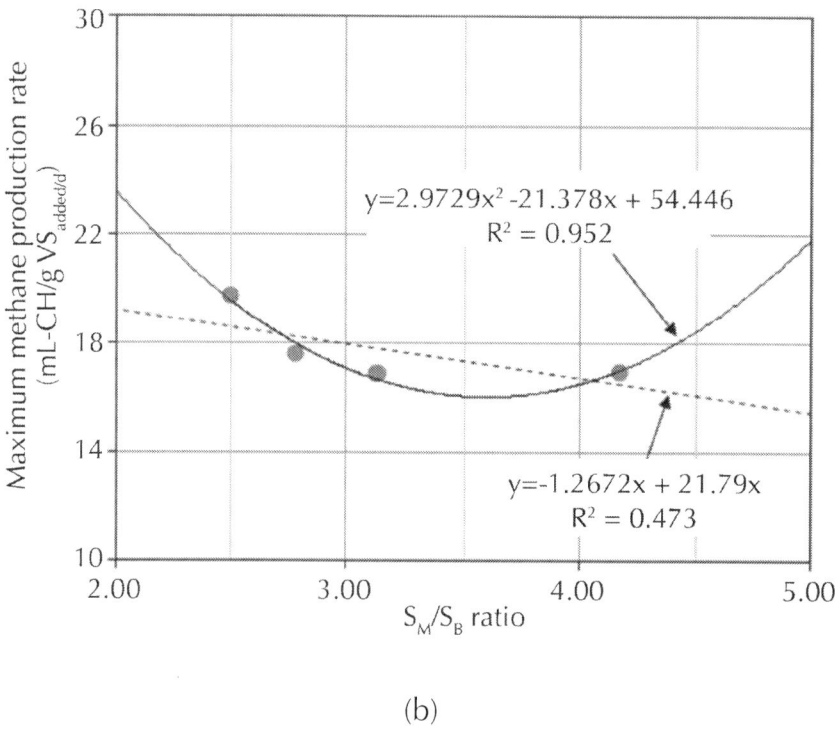

(b)

Figure 5: Change of methane yield coefficients in terms of M/H ratio. a) Ultimate methane yield; b) Maximum methane production rate.

CONCLUSIONS

Considering domestic waste and natural water management, piggery manure is mostly required to be treated among non-point sources of the stock raising system and can also be a main compound of methane production. With that background, this study performed a series of experiments on mesophilic-dry anaerobic digestion of the main compound, piggery manure, mixed with chaff, a mixture compound or substrate, for supplying carbon and controlling moisture. Also, this study conducted comparative evaluation of methane production potential produced from the anaerobic digestion process due to a mixing ratio of the piggery manure and the chaff, and obtained following conclusions after the careful consideration and emphasis on the reasonable mixing ratio:

- The larger amounts of the chaff resulted in the larger methane production yields as well as the larger gas production yields per volatile solid (VS) inserted. The methane ratio in the produced biogas was also increased as the amount of the chaff was increased.
- When using the same mixture ratio of the chaff, the gas production from the sample without trace elements was less than that from the sample with trace elements. Therefore, one can say that the trace elements are important when producing the methane from the mixture compounds of the piggery manure and the chaff.
- For the mesophilic-dry digestion, the cumulative methane production yield per the inserted VS was ranged in 163.02 mL - 313.27 mL-CH4/g-VS$_{added}$. Similar to the measurement result of total gas production yield, the methane production yield was increased with the higher mixture ratio of the chaff.
- For the comparative evaluation of the BMP assay, the ultimate methane yield using Modified Gompertz Model was calculated as 168.39 - 314.08 mL-CH$_4$/gVS$_{added}$. It was slightly lower than that in the thermophilic-wet digestion process for the mixture of the domestic food waste and livestock manure and similar to that in the thermophilic-wet digestion process of the sewage activated sludge.
- When the optimal mixture ratio (SM/SB) between the piggery manure (SM) and the chaff (SB) was less than 3.5, the ultimate methane production yield (M_o) and the maximum production rate (R_m) were decreased.

In conclusion, the mixture of the chaff to the piggery manure controls the moisture content and improves the methane production rate. Also, this study verifies the feasibility of the mesophilic-dry anaerobic digestion process, which is less difficult in operating and maintaining.

ACKNOWLEDGMENTS

This study was supported by a grant (CUHBRI-2012-0 2-007) of the CNUH-BRI as the project, Development of Future Oriented Healthcare Model and Fundamental Technology for Agro-Medical System, and partly supported by a grant (2013 research project) of Jeonbuk Green Environment Center. Also, we appreciate to Dr. Kyung-Yub Hwang (Research Fellow Emeritus) in Korea Institute of Science and Technology who gives us professional advice to complete this paper and experiments.

REFERENCES

1. P. Börjesson and B. Mattiasson, "Biogas as a ResourceEfficient Vehicle Fuel," Trends Biotechnology, Vol. 26, No. 1, 2008, pp. 7-13. http://dx.doi.org/10.1016/j.tibtech.2007.09.007
2. H. J. Park, M. K. Song and C. K. Na, "Pretreatment Efficiency of Piggery Wastewater Using Coagulation-MAP Sedimentation," Journal of Korea Society of Waste Management, Vol. 27, 2010, pp. 457-466.
3. Y. M. Yoon, Y. J. Kim and C. H. Kim, "The Evaluation of Economical Efficiency to Composting and Liquefying Process of Biomass Discharged in Pig Breeding," Agriculture Economics, Vol. 31, 2009, pp. 39-62.
4. D. Bolzonella, L. Innocenti, P. Pavan, P. Traverso and F. Cecchi, "Semi-Dry Thermophilic Anaerobic Digestion of the Organic Fraction of Municipal Solid Waste: Focusing on the Start-Up Phase," Bioresource Technology, Vol. 86, No. 2, 2003, pp. 123-129. http://dx.doi.org/10.1016/S0960-8524(02)00161-X
5. P. Pavan, P. Battistoni and J. Mata-Alvarez, "Performance of Thermophilic Semi-Dry Anaerobic Digestion Process Changing the Feed Biodegradability," Water Science and Technology, Vol. 41, 2000, pp. 75-81.
6. N. Forster-Carneiro, M. Perez and L. I. Romero, "Anaerobic Digestion of Municipal Solid Wastes: Dry Thermophilic

Performance," Bioresource Technology, Vol. 99, No. 17, 2008, pp. 8180-8184. http://dx.doi.org/10.1016/j.biortech.2008.03.021

7. B. Montero, J. L. Garcia-Morales, D. Sales and R. Solera, "Analysis of Methanogenic Activity in a Thermophilicdry Anaerobic Reactor: Use of Fluorescent in Situ Hybridization," Waste Management, Vol. 29, No. 3, 2009, pp. 1144-1151. http://dx.doi.org/10.1016/j.wasman.2008.08.010

8. S. E. Oh, M. K. Lee and D. H. Kim, "Continuous Mesophilic-Dry Anaerobic Digestion of organic Solid Waste," Journal of Korean Society of Environmental Engineers, Vol. 31, 2009, pp. 341-345.

9. J. J. Lay, Y. Y. Li and T. Noike, "Development of Bacterial Population and Methanogenic Activity in a Laboratory-Scale Landfill Bioreactor," Water Research, Vol. 32, No. 12, 1998, pp. 3673-3679. http://dx.doi.org/10.1016/S0043-1354(98)00137-7

10. W. F. Owen, D. C. Stuckey, J. B. Healy, L. Y. Young and P. L. McCarty, "Bioassay for Monitoring Biochemical Methane Potential and Anaerobic Toxicity," Water Research, Vol. 13, No. 6, 1979, pp. 485-492. http://dx.doi.org/10.1016/0043-1354(79)90043-5

11. R. S. Daniel and J. M. Tiedje, "General Method for Determining Anaerobic Biodegradation Potential," Applied and Environmental Microbiology, Vol. 47, 1984, pp. 850-857.

12. I. Angelidaki, M. Alves, D. Bolzonella, L. Borzacconi, J. L. Campos, A. J. Guwy, S. Kaalyuzhnyi, P. Jenicek and J. B. van Lier, "Defining the Biomethane Potential (BMP) of Solid Organic Wastes and Energy Crops: A Proposed Protocol for Batch Assays," Water Science and Technology, Vol. 59, No. 5, 2009, pp. 927-934.http://dx.doi.org/10.2166/wst.2009.040

13. M. H. Zwietering, I. Jongenburger, F. M. Rombouts and K. van't Riet, "Modeling of the Bacterial Growth Curve," Applied and Environmental Microbiology, Vol. 56, No. 6, 1990, pp. 1875-1881.

14. T. L. Hansen, J. E. Schmidt, I. Angelidaki, E. Marca, J. Cour Jansen, H. Mosboek and T. H. Christensen, "Method for Determination of Methane Potentials of Solid Organic Waste," Waste Management, Vol. 24, No. 4, 2004, pp. 393-400. http://dx.doi.org/10.1016/j.wasman.2003.09.009
15. H. B. Møllera, S. G. Sommera and B. K. Ahringb, "Methane Productivity of Manure, Straw and Solid Fractions of Manure," Biomass and Bioenergy, Vol. 26, No. 5, 2004, pp. 485-495. http://dx.doi.org/10.1016/j.biombioe.2003.08.008
16. J. K. Park, S. R. Jeong, J. H. Kang, Y. M. Ahn, H. E. Jin and N. H. Lee, "A Study on Optimization Condition for Anaerobic Co-Digestion of Food Waste with Livestock Wastes," Journal of Korea Society of Waste Management, Vol. 29, 2012, pp. 356-364.
17. S. H. Kim, H. C. Kim, C. H. Kim and Y. M. Yoon, "The Measurement of Biochemical Methane Potential in the Several Organic Waste Resources," Korean Journal of Soil Science and Fertilizer, Vol. 43, 2010, pp. 356-362.
18. S. Aslanzadeh, M. J. Taherzadeh and I. S. Horvath, "Pretreatment of Straw Fraction of Manure for Improved Biogas Production," Bio-Resources, Vol. 6, 2011, pp. 5193-5205.
19. J. G. Lin, Y. S. Ma, A. C. Chao and C. L. Huang, "BMP Tests on Chemically Pretreated Sludge," Bioresources Technology, Vol. 68, No. 2, 1999, pp. 187-192. http://dx.doi.org/10.1016/S0960-8524(98)00126-6
20. P. Shanmugam and N. J. Horan, "Simple and Rapid Methods to Evaluate Methane Potential and Biomass Yield for a Range of Mixed Solid Wastes," Bioresource Technology, Vol. 100, No. 1, 2008, pp. 471-474. http://dx.doi.org/10.1016/j.biortech.2008.06.027
21. D. Jackowiak, D. Bassard, A. Pauss and T. Ribeiro, "Optimization of a Microwave Pretreatment of Wheat Straw for Methane Production," Bioresource Technology, Vol. 102,

No. 12, 2011, pp. 6750-6756. http://dx.doi.org/10.1016/j.biortech.2011.03.107

22. J. C Converse, R. E. Graves and G. W. Evans, "Anaerobic Degradation of Dairy Manure under Mesophilic and Thernophilic Temperatures," Transactions of the ASAE, Vol. 20, 1977, pp. 336-340.

23. J. E. Robbins, M. T. Armold and S. L. Lacher, "Methane Production from Cattle Waste and Delignified Strawt," Infection and Immunity, Vol. 38, 1979, pp. 175-177.

Chapter 6

Hydrodynamics of Liquid Film in Helical Tubes

Mohammed Salah Hameed and Masab Kadhim Jawad

Chemical Engineering Department, Higher Colleges of Technology, Sharjah, United Arab Emirates

ABSTRACT

Hydrodynamic experiments on a liquid film are carried out using water in both straight and helical tubes at angles of inclination ranging between 2.5° and 5° and on three different coil diameters (23.86 cm, 32.74 cm and 41.13 cm) for film Reynolds numbers ranging from 100 to 2000. The film thickness is measured by two micrometers, arranged to measure vertical and horizontal distances within the cross section of the tube. The results of film thickness are related to the hydraulic radius to characterize the film flow in both types of tube. Momentum transfer rates are shown to be higher in helical tubes than in the straight incline tube. An empirical correlation is presented for film thickness in the helical tube in terms of N_T (coil tube)/N_T (straight tube) for film Dean number ranging from 1 to 1000.

INTRODUCTION

The occurrence and applications of film flow in chemical engineering processes are numerous; among them are absorption, extraction, heat transfer, humidification and distillation. The Wetted-wall column, a well know film flow technique, is frequently used for the experimental determination of mass transfer coefficients.

The presence of curves or bends is unavoidable in the design of open channel [1] thus producing spiral current and cross-waves in addition to the unique features of super elevation produced by centrifugal force.

Throughout the last century research work was mainly focused on film flow over flat surfaces or channels and over vertical tubes. Since 1960, extensive literature has been published dealing with wavy gas-liquid interfaces and concentrating mainly on the conditions under which waves exist and their effect on the processes of heat, mass and momentum transfer.

Hopf (cited in [2]) conducted experiments in a rectangular channel of slope ranging from 0.5° to 3.5° and noted the influence of turbulence between the critical film Reynolds numbers ($Re_{f,crit}$) ranging from 250 to 300. He found that wall roughness has no effect on $Re_{f,crit}$ except for the smallest depths. Nusselt (cited in [3]) gave a theoretical treatment for smooth, laminar and two dimensional film flows and stated that:

$$\delta = \left(\frac{3\upsilon^2}{g\sin\psi}\right)^{1/3} (\text{Re}_f)^{1/3} \qquad (1)$$

$$U_{av} = \frac{g\sin\psi}{3\upsilon}\delta^2 \qquad (2)$$

$$\frac{U_s}{U_{av}} = 1.5. \qquad (3)$$

$$f = \frac{6}{\text{Re}_f} \qquad (4)$$

$$N_T = (3\,\text{Re}_f)^{1/3} \qquad (5)$$

Jefferys [4] conducted experiments in channels at small slopes for a large range of Re_i. He confirmed the applicability of Equation (3) for the laminar region while the ratio of velocities decreases to 1.06 for the turbulent region. Cooper and Willey [5] determined experimental data for dilute sulfuric acid inside vertical tube and together with other workers found that up to Re_i equal to 350, the data are in excellent agreement with Equation (3). While Kirkbride's data [6] on film flowing outside a vertical tube deviated positively from the theoretical film thickness.

Fiend (cited in [2]) gave an experimental correlation for film flow of water and aqueous solutions with counter-current air in vertical tubes in turbulent region ($\text{Re}_i > 400$) as:

$$N_T = 0.369\left(3\operatorname{Re}_f\right)^{1/3} \tag{6}$$

Brauer (cited in [2]) gave an empirical equation for turbulent flow as:

$$N_T = 3^{1/3} \operatorname{Re}_f^{8/15} 400^{-1/5} \tag{7}$$

while Fulford [2] related the falling film thickness by dimensional analysis for channel flow, of slopes between 7.5° - 9.0°, over the range $30 < \operatorname{Re}_i < 300$, in the form:

$$N_T = 1.28(\sin\psi)^{-0.065} \operatorname{Re}_f^{0.331} \tag{8}$$

The equations mentioned above can be applied for quasi parallel-sided film i.e. for small film thickness flow inside or outside vertical tubes. While in straight inclined tubes and spiral tubes larger film thickness is expected but they differ in effect of centrifugal force on film in spiral tubes.

Most of the recent published works on helical pipes were mainly related to fully filled fluid flow in pipes and hardly any work was found dealing with film flow.

The latest work is concerned with the pipe flow belonging to Yamamoto and co-workers [7-9]. They studied the laminar and turbulent flow through helical coils. Their numerical and experimental data concluded a negligible effect of torsion on the flow within the range of their experimental data.

Several other workers investigated the emergence of turbulence region in flow through coiled pipes, both experimentally [8-11] and numerically [12, 13]. The coil curvature seems to increase the value of the Reynolds number required to attain fully turbulence flow. Difficulties were experienced in locating the transition region in the fully filled flow in helical pipes.

Vashisth et al. [14] published an intensive review on the performance of curved tubes for heat and mass transfer for fully filled flow.

Gupta et al. [15] studied the effect of coil pitch and coil diameter on the friction factor for five types of fully filled coils with different radii variation developed for Newtonian fluid. They formulated an empirical Equation (9) based on their experimental data.

$$f = \left(\frac{16}{\text{Re}}\right)\left(1 + aN_{Ge}^b\right) \tag{9}$$

where a & b are constants and N_{Ge} is the Germano number that is equal to Re multiplied by coil curvature.

$$\text{The Coil Curvature} = \frac{\pi^2 (D_c/d_t)}{\left[\pi(D_c/d_t)\right]^2 + (p/d_t)^2}$$

They concluded that Equation (9) predicts the observed coil friction factor values to within ±10%.

Film flow in curved tubes has not been given any attention in recent literature. The present work studies the thin film flow in curved tubes and extends its findings to large film thickness. The investigation covers experimental and theoretical film flow in both straight inclined and spiral tubes and how the two systems can be related. The work also discusses the case of film flow over flat surfaces.

EXPERIMENTAL DESIGN

The apparatus used in this work is shown schematically in Figure 1, the system consists mainly of a constant head tank A, transparent flexible tube B, storage tank C and centrifugal pump D to circulate the distilled water in the system.

The testing part is made of flexible tube B having an elliptic cross section with an inside minor axis of 1.510 cm and inside major axis of 1.888 cm and a thickness of 0.412 cm.

It is known from literature that films are practically smooth for angles of inclination less than 5° over a wide range of Re_i and would have limited ripples at higher range of Re_i. For this purpose, angles of inclination between 2.5° - 5° were studied.

In straight inclined tube experiments, the flexible tube is mounted on a flat steel plate. The far end of the tube away from the entrance is supported on a horizontal stand. It can be moved up or down to a suitable distance that can be conveniently measured by a traveling microscope in order to set the required include angle (2.5°, 3°,....5°).

The length of the developing region was estimated to be about 20 film thickness [16].

Hence, the film thickness is measured at 50 cm distance away from the entrance of the tube to insure that all measurements are made in the fully developed region.

A special micrometer, as shown in Figure 2, is used to measure the film thickness through holes in upper part of the flexible tube B. It consists of two micrometers arranged so that the vertical and horizontal distances of the cross sectional area of the flow can be measured. The whole arrangement is supported by a vertical stand, which allows the two micrometers to move to the required position near the system.

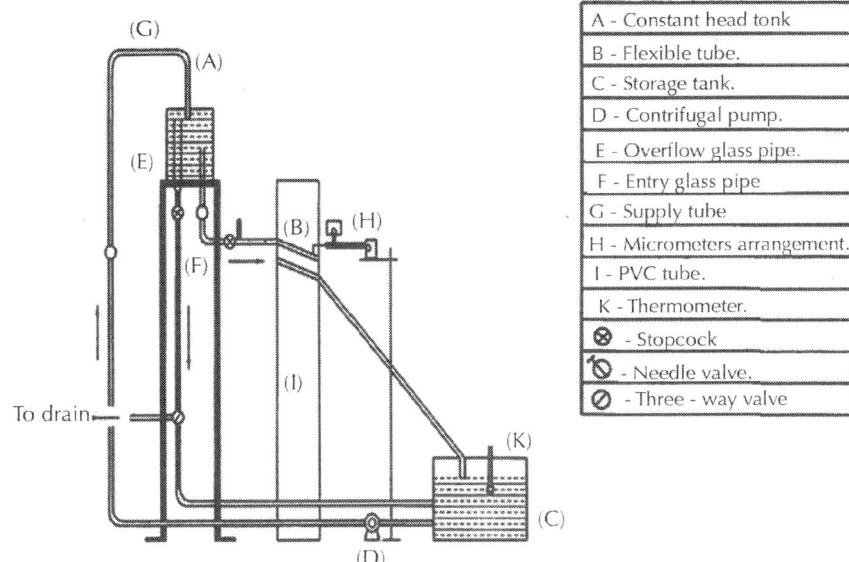

Figure 1: Chematic diagram of the apparatus used in film thickness measurement.

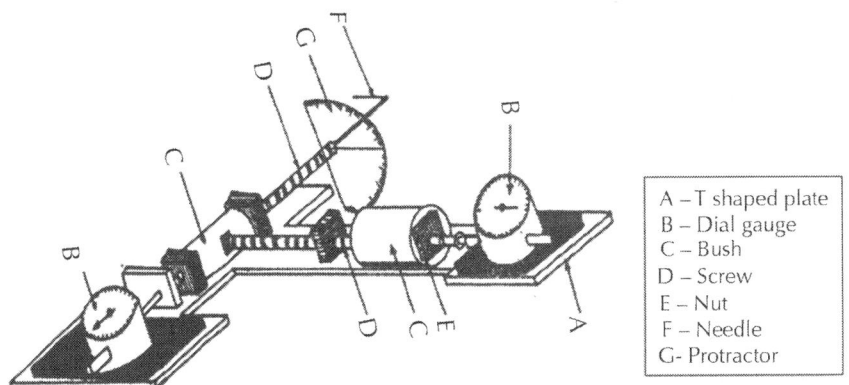

Figure 2: Micrometers arrangement.

In spiral tube experiments, the helical is made up by wrapping the flexible tube B around the PVC tube I. Three tubes I of different diameters are selected. The flexible tube B is wrapped at the same angles of inclination as mentioned above. The curvature of the tube has a calming action, the developing region in curved tubes is expected to be shorter than that in straight tubes.

THEORY AND MODEL

Film Flow in Inclined Circular Tube

The cross section of circular tube and a liquid film are shown in Figure 3. It can be obtained from trigonometric relations that:

$$b = a \frac{\cos\left(\dfrac{\alpha}{2}\right)}{\cos\left(\theta - \dfrac{\alpha}{2}\right)}$$

(10)

The Navier-Stokes equation in cylindrical coordinates for smooth, steady, laminar film flow can be reduced to the following equation:

$$\frac{d^2 U_z}{dr^2} + \frac{1}{r}\frac{dU_z}{dr} = -\frac{g \sin\psi}{\upsilon}$$

(11)

This differential equation can be solved by using the following boundary conditions:

Hydrodynamics of Liquid Film in Helical Tubes

$$\left.\begin{array}{l} \text{B. C. 1}: U_z = 0 \quad \text{at } r = a \\ \text{B. C. 2}: dU_z/dr = 0 \quad \text{at } r = b \end{array}\right] \text{ at all } \theta$$

The solution below represents the known semi-parabolic relation in the r-direction.

$$U_z = \frac{g \sin \psi a^2}{4\upsilon} \left[1 - \left(\frac{r}{a}\right)^2 + 2\left(\frac{b}{a}\right)^2 \ln\left(\frac{r}{a}\right) \right]. \tag{12a}$$

while the film surface velocity is represented by:

$$U_z|_s = \frac{g \sin \psi a^2}{4\upsilon} \left[1 - \left(\frac{b}{a}\right)^2 + 2\left(\frac{b}{a}\right)^2 \ln\left(\frac{b}{a}\right) \right]. \tag{12b}$$

The average velocity is obtained from the following relationship:

$$U_{a\upsilon} = \frac{\int_0^\alpha \int_b^a U_z r \, dr \, d\theta}{\int_0^\alpha \int_b^a r \, dr \, d\theta} \tag{13}$$

Submitting the expression of U_z in Equation (13) and carrying out the integration and then using the trigonometric identities would result in the following equation:

$$U_{av} = \frac{g\sin\psi a^2}{4v}\left[\frac{2}{a^2(\alpha-\sin\alpha)}\left(\frac{a^2}{4}\right)\right]$$
$$\cdot\left\{\alpha - \left(\frac{16}{3}\right)\left(\frac{\alpha}{2}\right)\cos^4\left(\frac{\alpha}{2}\right)\right.$$
$$\left. - \left[2\cos\left(\frac{\alpha}{2}\right)\sin\left(\frac{\alpha}{2}\right)\left(\frac{23}{9} - \frac{38}{9}\cos^2\left(\frac{\alpha}{2}\right)\right)\right]\right\} \quad (14)$$

Since $a^2(\alpha - \sin\alpha)/2$ represent the cross-sectional area of the film, the volumetric flow rate can be given as:

$$Q = \frac{g\sin\psi a^2}{4v}\left(\frac{a^2}{4}\right)\left\{\alpha - \left(\frac{16}{3}\right)\left(\frac{\alpha}{2}\right)\cos^4\left(\frac{\alpha}{2}\right)\right.$$
$$\left. - \left[2\cos\left(\frac{\alpha}{2}\right)\sin\left(\frac{\alpha}{2}\right)\left(\frac{23}{9} - \frac{38}{9}\cos^2\left(\frac{\alpha}{2}\right)\right)\right]\right\} \quad (15)$$

By comparing Equation (2), which is for the average velocity in the conventional two dimensional film flows with the derived Equation (14) which is for the average velocity in the inclined tube, it can be considered from their identity that the film thickness in inclined tube is defined by Equation (16):

$$\delta^2 = \frac{a^2}{2(\alpha-\sin\alpha)}\left\{\alpha - \left(\frac{16}{3}\right)\left(\frac{\alpha}{2}\right)\cos^4\left(\frac{\alpha}{2}\right)\right.$$
$$\left. - \left[2\cos\left(\frac{\alpha}{2}\right)\sin\left(\frac{\alpha}{2}\right)\left(\frac{23}{9} - \frac{38}{9}\cos^2\left(\frac{\alpha}{2}\right)\right)\right]\right\} \quad (16)$$

It is obvious that the film thickness is a function of both radius and central angle of the flow. Hence, the average velocity can be expressed as:

$$U_{av} = \frac{g \sin \psi}{4\nu} \delta^2 \qquad (17)$$

Further rearrangement of Equation (17) in terms of the dimensionless Nusselt film thickness and film Reynolds number would results in:

$$N_T = \left(4 \operatorname{Re}_f\right)^{1/3} \qquad (18)$$

As would be expected the flow in inclined circular tubes (Equation (18)) predicts higher Nusselt film thickness than in case of a flow over inclined flat plate (Equation (5)); since, the geometrical shape of the former offers higher wetted perimeter ($H_R/H_D < 1$) than in the latter case ($H_R/H_D = 1$) for the same free (exposed) surface.

It is more convenient to manipulate the experimental data for the flow in elliptic cross-section in terms of hydraulic radius rather than film thickness.

Applying the least squares fit to Equation (14) and the definition of hydraulic radius (refer to Figure 4) gives.

$$U_{av} = \frac{g \sin \psi}{4\nu} \left(2.0925 H_R^2\right) \qquad (19)$$

with a standard deviation of 4.578×10^{-3} and 0.94% average percentage error:

where

$$\delta = 1.4465 H_R \qquad (20)$$

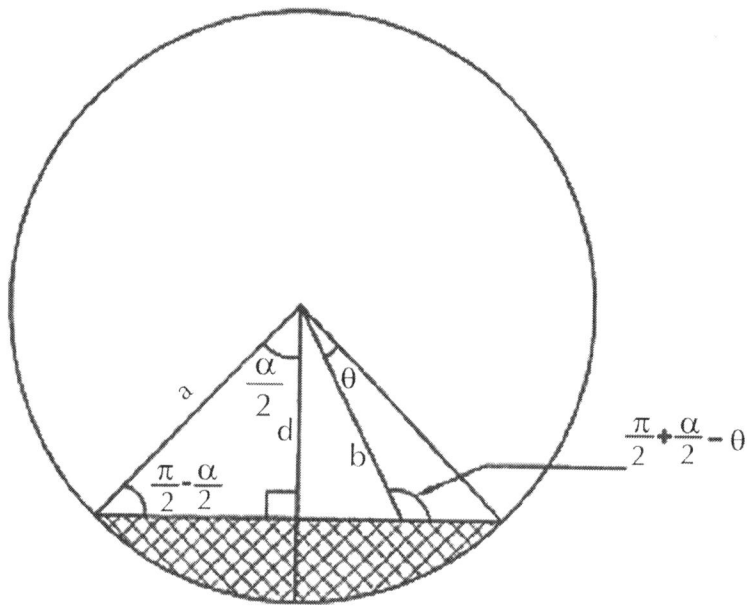

Figure 3: Liquid film in circular tube.

The Film Friction Factor

As the gas phase is practically stationary in the system, negligible drag force can be assumed at the free surface of the film. Therefore, the film weight is supported by the shear stress at the wall:

$$\tau_w = \delta \rho g \sin \psi \tag{21}$$

By substituting the value of d from Equation (17) and using the definition of Re_f, Equation (21) will be:

$$\tau_w = \rho\left(4v^2 g^2 \sin^2\psi\right)^{1/3} \operatorname{Re}_f^{1/3} \tag{22}$$

By defining the friction factor in the following form:

$$\tau_w = f\rho \frac{U_{av}^2}{2} \tag{23}$$

and using the expression for U_{av} from Equation (17), the following equation is obtained:

$$f\left(\frac{\rho}{2}\right)\left(\frac{vg\sin\psi}{4}\right)^{2/3} \operatorname{Re}_f^{4/3} = \rho\left(4v^2 g^2 \sin^2\psi\right)^{1/3} \operatorname{Re}_f^{1/3} \tag{24}$$

Simplifying the above equation will give a correlation relating the film friction factor and film Reynolds number:

$$f = \frac{8}{\operatorname{Re}_f} \tag{25}$$

This is similar to the equation for laminar flow in a closed pipe. Equation (25) predicts a higher friction factor than would be predicted for liquid film falling on an inclined flat plate. Here again, the difference is due to the higher wetted perimeter for circular tubes than that of flat plates for the same free surface which would result in a higher drag force.

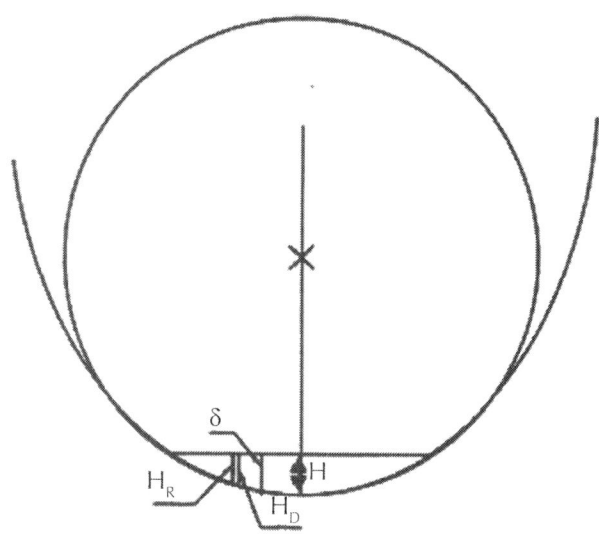

Figure 4: Hydraulic depth, hydraulic radius and film thickness at $Re_f = 303$, angle of inclination 3°.

RESULTS AND DISCUSSION

Inclined Tube

Figure 5 shows the variation of Nusselt film thickness versus film Reynolds number. Results show that laminar and turbulent regions fall within Re_f ranging from 480 to 600. The transition region is reconfirmed at about the same Re_f in Figure 6 in which the film friction factor, calculated from Equations (21) and (23), is plotted against the film Reynolds number.

In the laminar region ($Re_f < 480$) (see Figure 5), most of the experimental points lie above the theoretical line of Equation (18) with the maximum deviation falling between +30% to –7%. The least-squares fit of the experimental data gives the empirical equation:

$$N_T = 1.153(4\,\text{Re}_f)^{1/3} \tag{26}$$

with standard deviation = 0.158 and average % error = 10.33.

While for the turbulent region ($\text{Re}_i > 600$), the following empirical equation is obtained:

$$N_T = 0.872\,\text{Re}_f^{8/17} \tag{27}$$

with standard deviation = 6.43×10^{-2} and average % error = 0.503.

Considering the whole experimental range of the film Reynolds number the following empirical equation is obtained:

$$N_T = 1.225\,\text{Re}_f^{5/12} \tag{28}$$

with standard deviation = 0.164 and average % error = 9.64.

The three above empirical equations, are shown in Figure 7 together with Nusselt, Fiend and Brauer equations for film flow over flat surfaces. For film flow in inclined circular tubes a higher Nusselt film thickness for both the laminar and turbulent regions is predicted.

Ripple inception is observed to occur at Re_f ranging from 260 to 300 within all range of the angles of inclinations used in the experimental runs. Thomas, et al. [17] observed a ripple inception at lower range of Re_f i.e. between 80 - 130 for the central region of the film in a wide channel. This difference in behavior may be due to the relatively narrow tube used in this study, where surface tension would have greater influence on the flow.

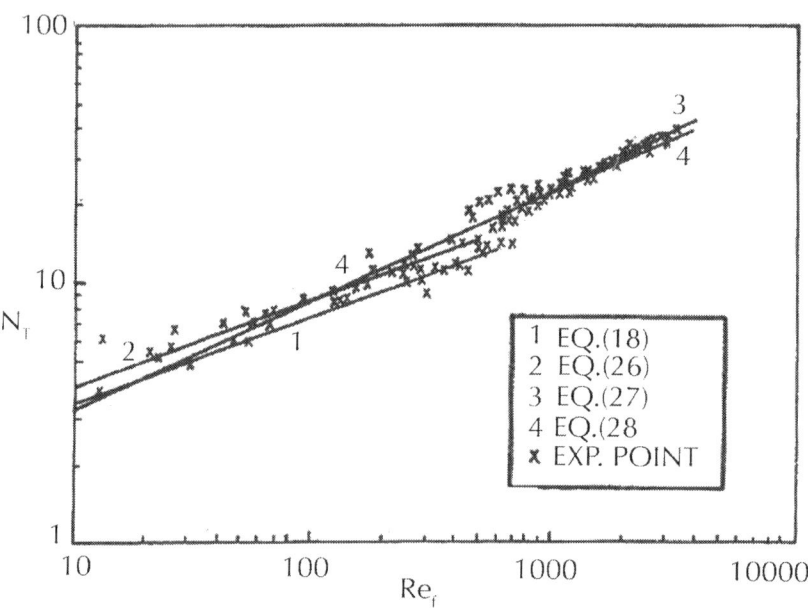

Figure 5: Inclined tube Nusselt number versus Reynolds number.

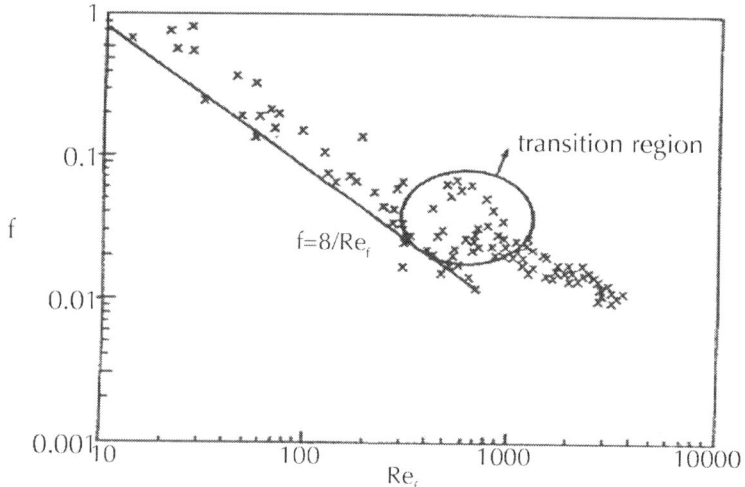

Figure 6: Film friction factor versus Reynolds number (inclined tube).

Helical Tube

A flexible transparent plastic tube is wrapped around a plastic cylinder at a predetermined angle of inclination, as in Figure 1. During the experimental runs, it was observed that the instability in film behavior increases with film flow rate and helix curvature.

As the flow rate is reduced, small cross waves were observed moving from the inner to the outer walls of the tube. These crosswaves disappear at lower flow rates and are replaced by small random ripples. At this stage four to five readings were recorded for film thickness.

First and second order polynomials were used to represent surface profile of the film and to extrapolate the surface shape near the inner and outer walls of the tube.

Most of the results suggested that first order polynomial correlation applied especially at low surface gradient. A typical representation of experimental runs is shown in Figure 8, which shows the unaxisymmetrical behavior of the film in helical tubes.

For each run the hydraulic radius, hydraulic depth and cross-sectional area of the flow are calculated. When the surface of the film was inclined, the surface length and the cross-sectional area of the flow were calculated using the analytic integration method for the surface equation.

In Figures 9 and 10, Nusselt film thickness is plotted against film Reynolds number. The transition region is not so distinguished as an inclined tube. In the helical tube gradual change of flow behavior confuses the limits of the transition region. This behavior is similar to itscounter part of a full pipe flow.

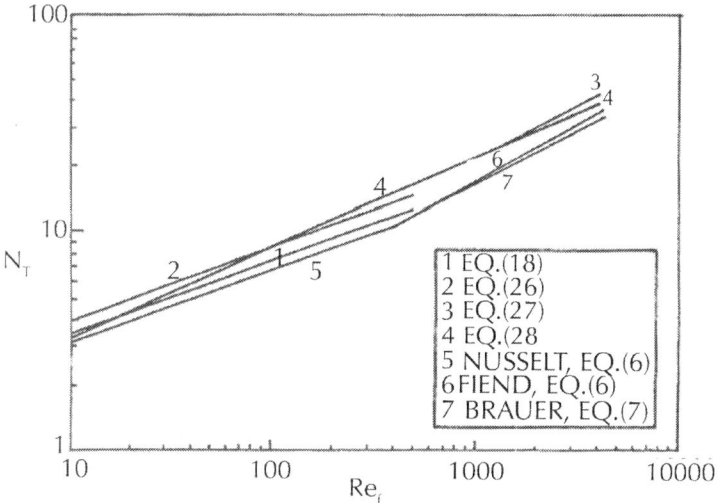

Figure 7: Comparison of present relation of film flow in inclined tube with that of inclined flat plate.

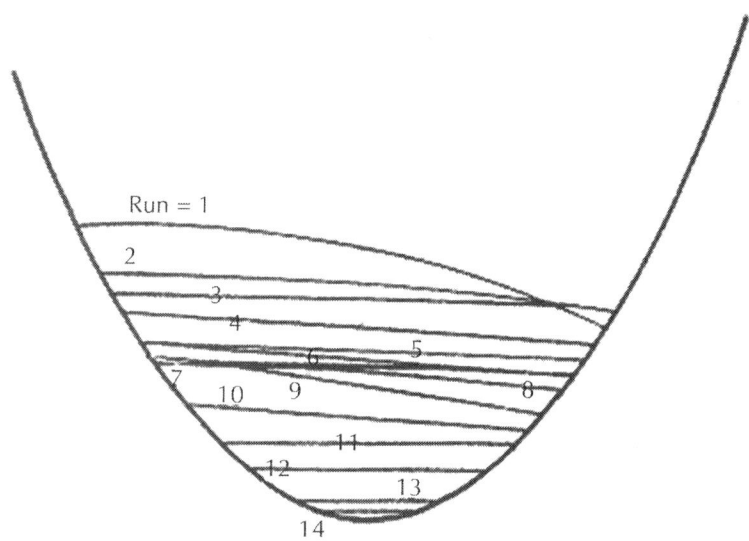

Figure 8: Film surface profile in elliptic cross sectional tube (polynomial approximation).

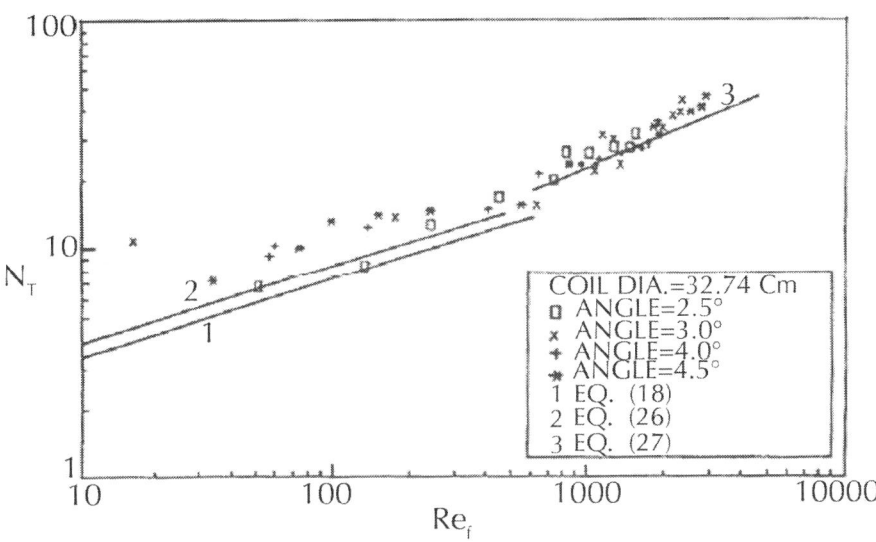

Figure 9: Helical tube Nusselt number versus Reynolds number (pitch effect at constant curvature).

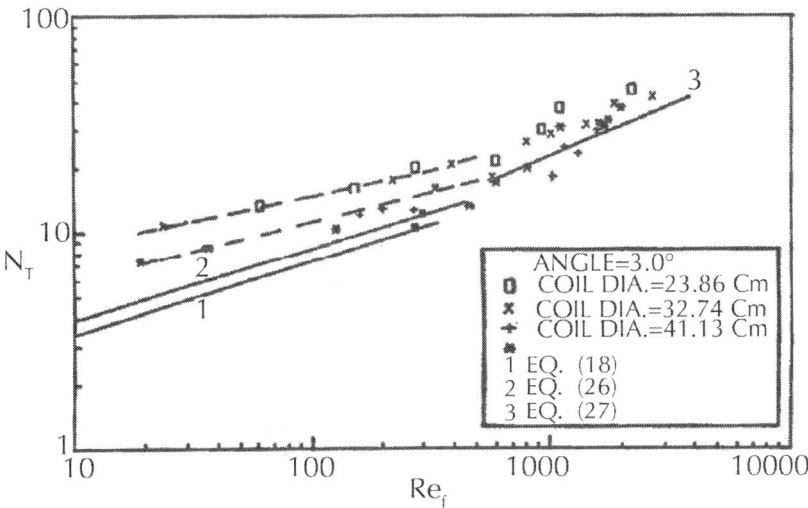

Figure 10: Helical tube Nusselt number versus Reynolds number (Curvature effect at constant pitch of 300 angle).

In the laminar region, the effect of pitch (angle of inclination) is clearly evident (see Figure 9) at certain film Reynolds number.

For turbulent flow, the action of coil diameter (curvature) and the action of centrifugal force would mostly reduce by inertia force and by the development of eddies. The effect of higher curvature can be observed while the pitch effect is overlapped by other effects.

In Figures 11 and 12, the Nusselt film thickness is plotted against the Dean number (De_i) which combines the effects of both curvature and film Reynolds number. The figures reveal that the spread in the experimental results is reduced as compared to Figures 9 and 10.

In order to relate Nusselt film thickness for coiled tubes to that of straight tubes, Figures 13 and 14 are plotted in terms of $N_{T(e.tube)} / N_{T(st.tube)}$ versus film Dean number. Equation (28) is applied to calculate $N_{T(st.tube)}$. It can be observed that most of the experimental points lie between value of $N_{T(e.tube)} / N_{T(st.tube)}$ of 1 - 2. This ratio tends to decrease gradually with the increase of De_i at certain coil diameter.

By the least squares fit, the following empirical correlation is obtained:

$$\frac{N_{T(c.tube)}}{N_{T(st.tube)}} = 11.0 De_f^{-0.1} \left(\frac{a}{R}\right)^{0.55}$$

(29)

with Standard Deviation = 0.106 and average % error = 9.56.

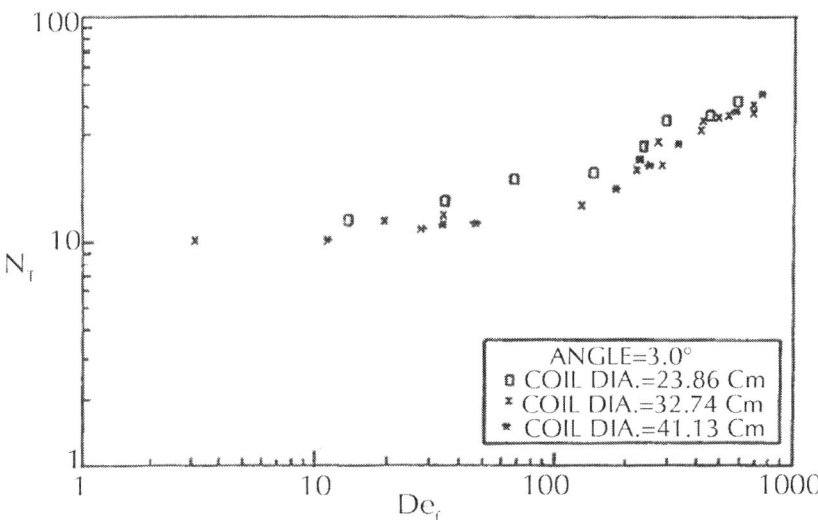

Figure 11: Helical tube Nusselt number versus Dean number (curvature effect at constant pitch of 3.00 angle).

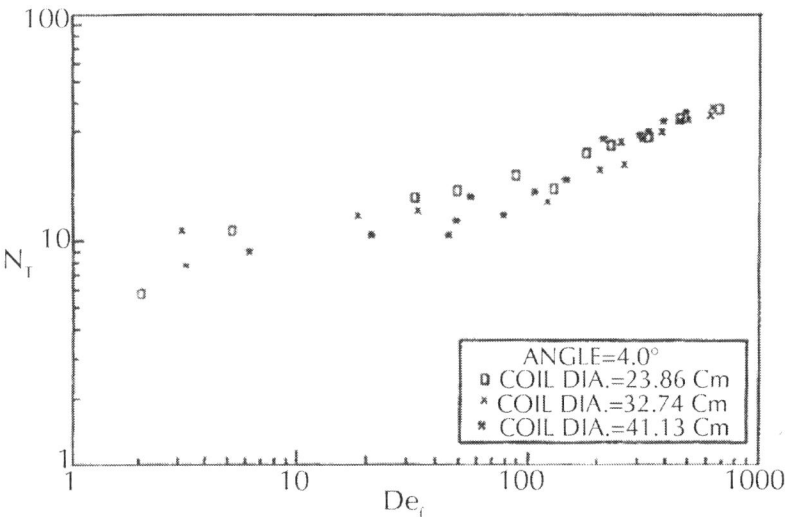

Figure 12: Helical tube Nusselt number versus Dean number (curvature effect at constant pitch of 4.00 angle).

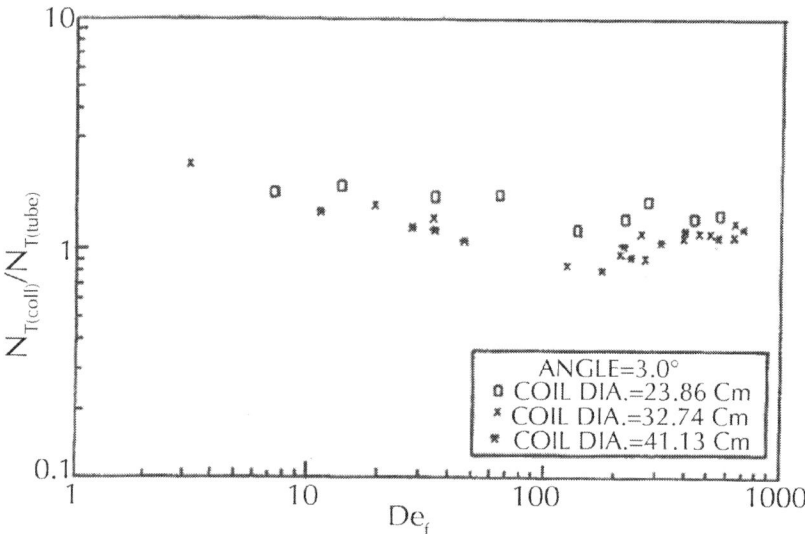

Figure 13: Ratio of Nusselt number of helical tube over Nusselt number (curvature effect at constant pitch of 3.00 angle).

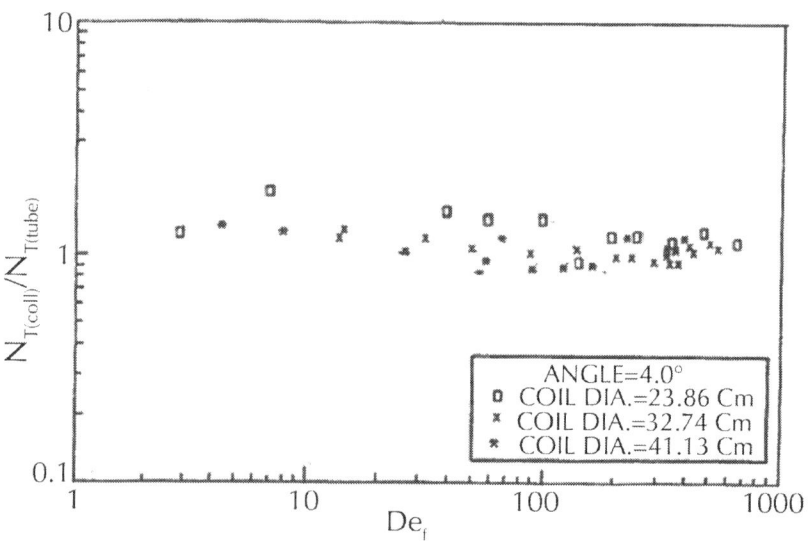

Figure 14: Ratio of Nusselt number of helical tube over Nusselt number (curvature effect at constant pitch of 4.00).

CONCLUSIONS

- Systems that have their film thickness vary peripherally can be characterized by some average film thickness. This is a function of hydraulic radius defined by the film boundary.
- In helical tubes, the film thickness increases with the increased curvature accompanied by a reduction in average velocity. Within the range of the obtained data, coiling effect may account for the maximum increase in film thickness of 70% over that of the inclined tube. But, this increase diminishes gradually in the turbulent region, where the secondary flow intensity is small relative to axial flow intensity.
- Pitch within two to three tube diameters in length has no appreciable effect on film thickness at low tube curvature.
- The transition region in a straight inclined tube is marked between Re_i = 480 - 600, which is higher than the reported range for vertical tubes (Re_i = 350 - 500). In helical tubes, the transition region is not so distinct (within experimental data) as it is in inclined tube 5) Results are correlated empirically for the cases of straight inclined and helical tubes. The former is expressed in terms of N_T as function of Re_i in the laminar and turbulent regions. While for the latter the empirical correlation is represented in terms of the ratio of N_T to coil over N_T for straight inclined tubes as function of De_f.

REFERENCES

1. V. T. Chov, "Open-Channel Hydraulic," McGraw-Hill, Kogakusha, 1959.
2. G. D. Fulford and T. B. Drew, "Advances in Chemical Engineering," Academic Press, New York, Vol. 5, 1964.
3. R. B. Bird, W. E. Steward and E. N. Lightfoot, "Transport Phenomena," John Wiley, New York, 1960.
4. H. Jefferys, "The Flow of Water in an Inclined Channel of Rectangular Section," Philosophical Magazine, Vol. 49, 1925.

5. C. M. Cooper, T. B. Drew and W. H. McAdams, Transactions of the American Institute of Chemical Engineers, Vol. 30, 1934.
6. C. G. Kirkbride, Industrial & Engineering Chemistry Research, Vol. 26, 1934.
7. K. Yamamoto, S. Yanase and T. Yoshida, "Torsion Effect on the Flow in a Helical Pipe," Fluid Dynamics Research, Vol. 14, 1994, pp. 259-273. doi:10.1016/0169-5983(94)90035-3
8. K. Yamamoto, T. Akita, H. Ikeuchi and Y. Kita, "Experimental Study of the Flow in a Helical Circular Tube," Fluid Dynamics Research, Vol. 16, No. 4, 1995, pp. 237-249. doi:10.1016/0169-5983(95)00022-6
9. K. Yamamoto, A. Aribowo, Y. Hayamizu, T. Hirose, K. Kawahara, "Visualization of the Flow in a Helical Pipe," Fluid Dynamics Research, Vol. 30, No. 4, 2002, pp. 251-267. doi:10.1016/S0169-5983(02)00043-6
10. D. R. Webster and J. A. C. Humphrey, "Experimental Observation of Flow Instability in a Helical Coil," Fluid Engineering, Vol. 115, 1993, pp. 436-443. doi:10.1115/1.2910157
11. A. Cioncolini and L. Santini, "An Experimental Investigation Regarding the Laminar to Turbulent Flow Tansition in Helically Coiled Pipes," Experimental Thermal and Fluid Science, Vol. 30, No. 7, 2006, pp. 367-380. doi:10.1016/j.expthermflusci.2005.08.005
12. T. J. Hüttl and R. Friedrich, "Influence of Curvature and Torsion on Turbulent flow in helically Coiled Pipes," International Journal of Heat and Fluid Flow, Vol. 21, No. 3, 2000, pp. 345-353.
13. T. J. Hüttl and R. Friedrich, "Direct Numerical Simulation of Turbulent Flows in Curved and helically Coiled Pipes," Computers & Fluids, Vol. 30, No. 5, 2001, pp. 591-605. doi:10.1016/S0045-7930(01)00008-1
14. V. Vashisth, V. Kumar and K. D. P. Nigam, "A Review on That Potential Applications of Curved Geometries in Process

Industry," Industrial & Engineering Chemistry Research, Vol. 47, No. 10, 2008, pp. 3291-3337. doi:10.1021/ie701760h
15. R. Gupta, R. K. Wanchoo and T. R. M. J. Ali, "Laminar Flow in Helical Coils: A Parametric Study," Industrial & Engineering Chemistry Research, Vol. 50, No. 2, 2011, pp. 1150-1157. doi:10.1021/ie101752z
16. B. Atkinson and R. L. McKee, Chemical Engineering Science, Vol. 19, 1964.
17. W. J. Thomas, M. S. Ray and E. W. Palmer, "Physical Absorption of Carbon Dioxide in Water Flowing in an Inclined Cell," Chemical Engineering Communications, Vol. 2, No. 3, 1976, pp. 121-134. doi:10.1080/00986447608960454

Chapter 7

The Functional Potential of Microbial Communities in Hydraulic Fracturing Source Water and Produced Water from Natural Gas Extraction Characterized by Metagenomic Sequencing

Arvind Murali Mohan[1], Kyle J. Bibby[2], Daniel Lipus[3], Richard W. Hammack[4], and Kelvin B. Gregory [4]

[1]National Energy Technology Laboratory, Pittsburgh, Pennsylvania, United States of America

[2]Department of Civil and Environmental Engineering, Carnegie Mellon University, Pittsburgh, Pennsylvania, United States of America

[3]Department of Civil and Environmental Engineering, University of Pittsburgh, Pittsburgh, Pennsylvania, United States of America

[4]Department of Computational and Systems Biology, University of Pittsburgh Medical School, Pittsburgh, Pennsylvania, United States of America

ABSTRACT

Microbial activity in produced water from hydraulic fracturing operations can lead to undesired environmental impacts and increase gas production costs. However, the metabolic profile of these microbial communities is not well understood. Here, for the first time, we present results from a shotgun metagenome of microbial communities in both hydraulic fracturing source water and wastewater produced by hydraulic fracturing. Taxonomic analyses showed an increase in anaerobic/facultative anaerobic classes related to *Clostridia*, *Gammaproteobacteria*,*Bacteroidia* and *Epsilonproteobacteria* in produced water as compared to predominantly aerobic *Alphaproteobacteria* in the fracturing source water. The metabolic profile revealed a relative increase in genes responsible for carbohydrate metabolism, respiration, sporulation and dormancy, iron acquisition and metabolism, stress response and sulfur metabolism in the produced water samples. These results suggest that microbial communities in produced water have an increased genetic ability to handle stress, which has significant implications for produced water management, such as disinfection.

INTRODUCTION

High-volume hydraulic fracturing operations for natural gas development from deep shale produce millions of gallons of wastewater over the lifetime of the well [1], [2], [3], commonly termed as 'produced water'. This produced water contains elevated concentrations of salts, metals, hydrocarbons and radioactive elements [3], [4], [5], [6], [7]. Microbial communities in produced water can utilize hydrocarbons as sources of carbon and energy [8] and transform redox labile salts and metals. This can give rise to significant water management challenges [9]and increased production costs [10], [11]. For instance, sulfidogenic and acid producing bacteria can cause corrosion of metal infrastructure, souring of natural gas, and reduced formation permeability [10], [11], [12], [13].

Deleterious microbial activity is commonly controlled with biocides at significant cost to the driller. However, despite biocide use, microbial activity is prevalent in produced water. Previous studies have shown that biocide effectiveness may be limited by high salt concentrations, organic compounds, and long residence times in the subsurface [14], [15], [16]. Other studies have shown that microbial communities in produced water are distinct from those in the injected fracturing fluid, and correlate well with changes in geochemical and environmental conditions[5], [15], [17]. This implies that the common practice of recycling produced water for subsequent hydraulic fracturing may introduce adapted populations into the formation [5].

Over the past decade molecular ecology surveys based on the 16S rRNA gene have increased our knowledge about the taxonomic composition of microbial communities in reservoir environments [5], [15], [17], [18], [19], [20], [21], [22]. However, these studies offer limited insights on the metabolic capabilities of the microbial community, as they rely on taxonomic inference based on 16S rRNA gene similarity to previously isolated microorganisms. As an example of the limitations of using previously isolated microorganisms to infer metabolic capability, the 'core genome' of the well-studied *Escherichia coli* is typically less than 50% of the genes in the genome, and <30% of the *E. coli* pan-genome [23]. On the other hand, shotgun metagenomic surveys enable access to complete genetic information within microbial genomes from uncultured, mixed consortia [24], [25], [26]. These surveys have provided significant insights on the functional potential of microorganisms in diverse environments such as marine samples [25], corals [27], activated sludge [28], permafrost [29], hydrocarbon and sandstone reservoirs [30], [31], and swine gut [32]. Despite the importance of microbial activity in produced water brines from hydraulic fracturing operations, the functional potential of associated microbial communities has not yet been studied. In this study, the metagenome of fracturing source water and produced water at two different time points from a Marcellus Shale natural gas well in Westmoreland County, PA was generated using Illumina MiSeq technology. The microbial ecology from 16S rRNA surveys and chemical composition of these samples has been described in a previous publication [5]. Sequences from each sample were assembled into contiguous sequences (contigs) and analyzed for taxonomic affiliations and functional potential of the microbial communities.

MATERIALS AND METHODS

Sampling

Samples of hydraulic fracturing source water, and produced water on days 1 and 9 were collected from a horizontally drilled Marcellus Shale natural gas well in Westmoreland County, Pennsylvania, U.S.A in October 2011. The source water used for fracturing was a mix of fresh reservoir water (~80%) and produced water (~20%) from previous fracturing operations. Fracturing additives amended to the source water included proppant (silica sand), scale inhibitor (ammonium chloride), biocide (mixture of tributyl tetradecyl phosphonium chloride, methanol and proprietary chemicals), hydrochloric acid, gel (paraffinic solvent), breaker (sodium persulfate) and friction reducer (hydrotreated petroleum distillate). Details regarding the sampling procedure and chemical additives used in the fracturing process are described elsewhere [5]. The aqueous geochemical characteristics of these samples were described previously [5].

DNA Extraction, Library Preparation and Illumina Sequencing

Unfiltered water samples were centrifuged at 6,000 g for 30 min in an Avanti J-E centrifuge (Beckman Coulter, Brea, CA) to pellet cells. DNA was extracted from 0.25 g of cell pellet using MO BIO power soil DNA isolation kit (MO BIO, Carlsbad, CA) according to the manufacturer's instructions. DNA was prepared using Nextera XT DNA sample preparation kit (Illumina, San Diego, CA) according to manifacturer's instructions at Genewiz (South Plainfield, NJ). DNA for sequencing was quantified using qPCR prior to clustering, and sequenced using the Illumina MiSeq (Illumina, San Diego, CA) with a 2×250 PE configuration at Genewiz, NJ. Sequencing demultiplexing was performed on the Illumina MiSeq instrument using sample-specific barcodes.

Bioinformatic Analyses

The raw unpaired sequences were checked for sequencing tags and adapters using the predict function implemented within the TagCleaner program [33]. No sequencing tags or adapters were identified. Sequences were then subjected to quality control using the FastX toolkit within the Galaxy platform [34] with a minimum length 100 and minimum quality score 20. The velvet assembler [35] was used to assemble sequences that passed quality control into contiguous sequences. The assembly parameters were empirically optimized for the dataset prior to assembly; the dataset was processed using a kmer length of 77. Generated contigs >500 bp in length were uploaded to the MG-RAST server [36] with associated metadata files for taxonomic affiliations and functional annotations. Sequence similarity searches in MG-RAST was performed using the BLAT tool [37]. The metagenomes from fracturing source water, day 1 produced water, and day 9 produced water are available in the MG-RAST server [36] under accession nos. 4525703.3, 4525704.3 and 4525705.3, respectively. Taxonmic assignments of selected funcional categories from MG-RAST were excecuted in MGTAXA [38], [39], on the Galaxy bioinformatics workbench [40], [34], using default parameters and taxonomy as defined by the NCBI taxonomic tree. Data is for contig abundance and does not reflect read mapping.

As an additional assembly-independent analysis, sequence data was mapped against reference genomes downloaded from NCBI with CLC Genomics Workbench (Version 6.5.1, CLC Bio, Aarhus, Denmark) [41] using default parameters and no masking. Reference genomes were selected based upon taxonomic observations in MG-RAST annotation and a previous microbial ecology investigation [5]. Prior to mapping, sequencing data was trimmed to a minimum length of 100 bp and minimum quality score of 20. Furthremore, sequences for the sulfite reductase subunits A and B (dsrA/dsrB) and the suflur metabolism gene adenylyl sulfate reductase subunit A (apsA) were downloaded from NCBI and mapped against the trimmed sequencing data using CLC Genomics Workbench (Version 6.5.1, CLC Bio, Arhus, Denmark).

RESULTS AND DISCUSSION

A total of 10 002, 17 055 and 16 661 contigs from the fracturing source water, produced water day 1 and day 9 samples, respectively, were uploaded to MG-RAST for downstream analyses. All uploaded contigs passed MG-RAST quality control and de-replication filters. The metagenomics sequence statistics are summarized in Table 1.

Table 1: Metagenomic sequence statistics of fracturing source water (SW), produced water day 1 (PW day 1) and produced water day 9 (PW day 9)

	SW	Pw day 1	PW day 9
Total base pair (bp) count	7,939,565 bp	18,254,354 bp	15,253,129 bp
No. of Contigs	10,002	17,055	16,661
Mean length of Contigs	793±809 bp	1,070±1,195 bp	915±651 bp

% GC content in Contigs	59±8%	55±13%	43±9%
% Contigs containing predicted proteins with known functions	83%	93.1%	80 8%
% Contigs containing predicted proteins with unknown functions	16.6%	6.6%	18.9%
% Contigs containing rRNA genes	0.4%	0.3%	0.3%
Identified protein features	9,919	20,687	16,982
Identified functional categories	8,041	16,948	13, 570

Taxonomic Composition

Taxonomic affiliations were assigned to contigs with predicted proteins and rRNA genes based on comparison with the M5NR database. Alpha diversity (predicted phylotypes) for the fracturing source water, produced water day 1 and day 9 samples were 90, 79 and 88, respectively. Rarefaction curves for each of the samples were asymptotic suggesting that the majority of taxonomic diversity was recovered from the samples. Alpha diversity values and rarefaction curves were obtained using the MG-RAST tool.

Bacteria constituted the dominant domain (97–99% of the total community) in all samples. However, a shift in bacterial community composition was detected between the samples at the class and order levels (Figure 1, 2). Contigs affiliated to the class *Alphaproteobacteria* constituted the majority of the community in the fracturing source water (81%) and produced water day 1 (67%) samples (Figure 1). Within *Alphaproteobacteria*, the dominant order detected was *Rhodobacterales* (68–88% of the *Alphaproteobacteria*; 55–59% of the total community) in both the source water and produced water day 1 samples (Figure 2). The relative abundance of *Alphaproteobacteria* decreased to <2% of the community in the produced water day 9 sample. Previous qPCR analysis of these samples suggests that that the total bacterial population remained constant at 10^6–10^7 copies of 16S RNA gene/ml [5].

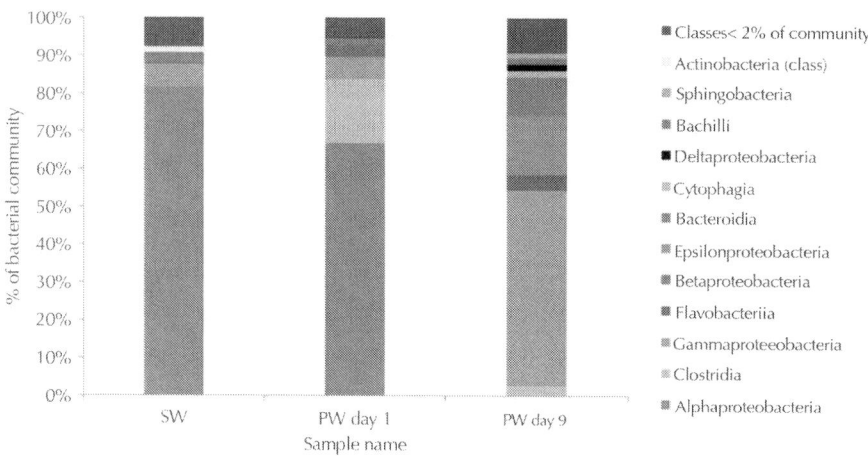

Figure 1: Class level affiliations assigned to contigs with predicted proteins and rRNA genes in source water (SW), produced water day 1 (PW day 1) and produced water day 9 (PW day 9).

Total community includes *Bacteria*, *Archaea*, *Viruses* and *Eukaryota*.

Order (class)	SW	PW day 1	PW day 9
Rhodobacterales (Alphaproteobacteria)			
Caulobacterales (Alphaproteobacteria)			
Rhizobiales (Alphaproteobacteria)			
Sphingomonadales (Alphaproteobacteria)			
Burkholderiales (Betaproteobacteria)			
Bacteroidales (Bacteroidia)			
Thermoanaerobacterales (Clostridia)			
Clostridiales (Clostridia)			
Flavobacteriales (Flavobacteria)			
Alteromanadales (Gammaproteobacteria)			
Vibrionales (Gammaproteobacteria)			
Enterobackteriales (Gammaproteobacteria)			
Cytophagales (Cytophagia)			
Campylobacterales (Epsilonproteobacteria)			

Color code	
2-5%	
>5-10%	
>10-20%	
>20-30%	
>30-40%	
>40-60%	

Figure 2: Order level affiliations assigned to contigs with predicted proteins and rRNA genes in source water (SW), produced water day 1 (PW day 1) and produced water day 9 (PW day 9).

Total community includes *Bacteria*, *Archaea*, *Viruses* and *Eukaryota*. Only orders representing >2% of the total community are shown in the figure.

An increase in the number of contigs associated with the class *Clostridia* was observed in the produced water day 1 sample (17%) as compared to the fracturing source water (1%). However, the relative abundance of *Clostridia* decreased to 3% in the produced water day 9 sample. The majority of the *Clostridia* in the produced water day 1 sample were affiliated to the order *Thermoanaerobacterales* (94% of *Clostridia*; 16% of the total community) (Figure 2).*Gammaproteobacteria* sequences constituted a minor fraction (6%) of the total community in the fracturing source water and produced water day 1 samples but increased in relative abundance to constitute the dominant class (52%) in the produced water day 9 sample. Within the *Gammaproteobacteria* of the produced water day 9 sample, dominant orders included*Vibrionales* (67% of *Gammaproteobacteria*) and *Alteromonadales* (23% of*Gammaproteobacteria*) (Figure 2). The day 9 samples also showed an increase in relative abundance of *Epsilonproteobacteria* (16%) and *Bacteroidia* (10%) classes as compared to the other samples (<2% of the total community). The major bacterial phyla, classes and orders identified in this study were consistent with previous 16S rRNA gene based clone library and pyrosequencing surveys of these samples [5]. These results indicate a shift towards facultative anaerobic/anaerobic and halophilic communities in the produced water samples as compared to a predominantly aerobic community in the fracturing source water. At the class level, in each of the samples less that 3% of the total sequences did not affiliate to any taxonomic group.

A minor fraction of the total community was represented by contigs affiliated to *Archaea* (0.1–0.4%), *Viruses* (0.3–1%) and *Eukaryota* (0.4–1.4%) domains. These domains were not analyzed for in the previous 16S rRNA gene survey of these samples [5], and were not considered in more detailed functional classification of the metagenomes.

Mapping Results

Metagenomic reads were mapped against a diverse set of reference genomes to confirm MG-RAST taxonomic results and only reference genomes with good mapping results are discussed in this section. Reference genome mapping results confirmed taxonomic MG-RAST contig analysis. The best mapping results for source water were obtained when sequences were mapped against reference genomes of *Alphaproteabacteria*, specifically of the order*Rhodobacterales* (Figures 3, 4). Similarly, produced water day 1 sample mapping results suggest that it was dominated by bacteria of the orders *Rhodobacterales* and*Thermoanaerobacterales* (Figures 3, 4). A distinct shift in bacterial community was observed between produced water day 1 samples and produced water day 9 samples based on mapping results. Best mapping results for produced water day 9 samples were obtained for reference genomes in the order *Campylobacterales* and *Alteromondales* further supporting the MG-RAST results (Figures 3, 4). Produced water samples demonstrated a distinctive signature with reads mapping best to few select reference genomes, while source water sample reads were distributed more evenly throughout all included reference genomes. For four reference genomes (*Thermoanaerobacter sp.* X514, *Thermoanaerobacter pseudethanolicus*,*Thermoanaerobact er mathranii* in produced water day 1 samples and *Marinobacter hydrocarbonoclasticus* DSM 7299 *in* produced water day 9 sample) more than 80% coverage was achieved suggesting that these species could play important roles in the microbial community of the representative sample (Figure 3). Highest observed reference genome coverage for source water sample sequences were 79% for *Roseovarius* sp. 217, 40% for*Ruegeria pomeroyi* and 38% for *Rhodobacter sphaeroides* (Figure 3). For produced water day 1 samples, about 10% of all trimmed sequencing reads mapped against the three*Thermanaerobacter* genomes included in the analysis and 8–13% of reads mapped successfully against *Roseovarius sp.* 217 and *Roseovarius nubinhibens* genomes (Figure 4). 7.7% of produced water day 1 reads mapped against the *Ruegeria pomeroyi* genome (Figure 4). 4–6% of reads for produced water day 9 samples mapped against two different*Marinobacter* and *Arcobacter* reference genomes and one *Vibrio* reference genome (Figure 4). Almost 16% of all reads from source water samples mapped against *Roseovarius sp.* 217 and approximately 4–6% of reads for source water sample mapped against each *Dinoroseobacter shibae, Ruegeria pomeroyi, Rhodobacter sphaeroides* and *Rhodobacter capsulatus* genomes (Figure 4). The high number of reads form source water and produced water day 1 samples mapping against *Roseovarius* species is in agreement with previous 16S rRNA gene sequencing [5], implying the *Roseovarius* species might be of importance in these waters. *Roseovarius sp.* was previously identified in natural gas brines from the Marcellus shale and its potential implications are discussed elsewhere [9].

Coverage of Reference Genomes	SW	PW day1	PW day9
Dinoroseobacter shibae	0.36	0.39	0.20
Thermoanaerbacter sp.	0.01	0.86	0.05
Thermoanaerbacter pseudenthanolicus	0.02	0.87	0.05
Ruegeria pomeroyi	0.40	0.43	0.23
Thermoanaerbacter tengcongensis	0.01	0.45	0.03
Roseobacter denitrificans	0.30	0.33	0.15
Flavobacterium psychrophilum	0.18	0.27	0.05
Arcobacter butzleri	0.31	0.52	0.47
Arcobacter nitrofigilis	0.23	0.50	0.39
Marinobacter hydrocarbonoclasticus	0.14	0.19	0.83
Bacteroides fragilis	0.02	0.06	0.13
Sulfospirillum deleyianum	0.05	0.07	0.08
Sulfurimonas denitrificans	0.15	0.13	0.12
Parabacteroides distasonis	0.02	0.07	0.10
Phenylobacterium zucineum	0.33	0.27	0.08
Jannaschia sp.	0.25	0.27	0.12
Rhodobacter sphaeroides	0.38	0.42	0.21
Hyphomonas neptunium	0.17	0.22	0.08
Flavobacterium johnsoniae	0.09	0.14	0.03
Rhodobacter capsulatus	0.34	0.37	0.18
Marinobacter adhaerens	0.14	0.17	0.68
Clostridium difficile	0.02	0.06	0.06
Roseovarius sp.	0.79	0.78	0.51
Roseovarius numbinhibens	0.47	0.50	0.28
Vibrio campbelli	0.06	0.25	0.51
Thermoanaerobacter mathranii	0.01	0.88	0.05

Color code
>0-0.1
>0.1-0.2
>0.2-0.3
>0.3-0.4
>0.4-0.5
>0.5-0.6
>0.6-0.7
>0.7-0.8
>0.8-0.9

Figure 3: Fraction of genome coverage for source water (SW), produced water day 1 (PW day 1) and produced water day 9 (PW day 9) samples.

Reads were mapped against reference genomes using CLC Genomic workbench version 6.5.1 using default parameters. Shown are fractions of reads mapped against each reference genome included in the analysis for all three samples.

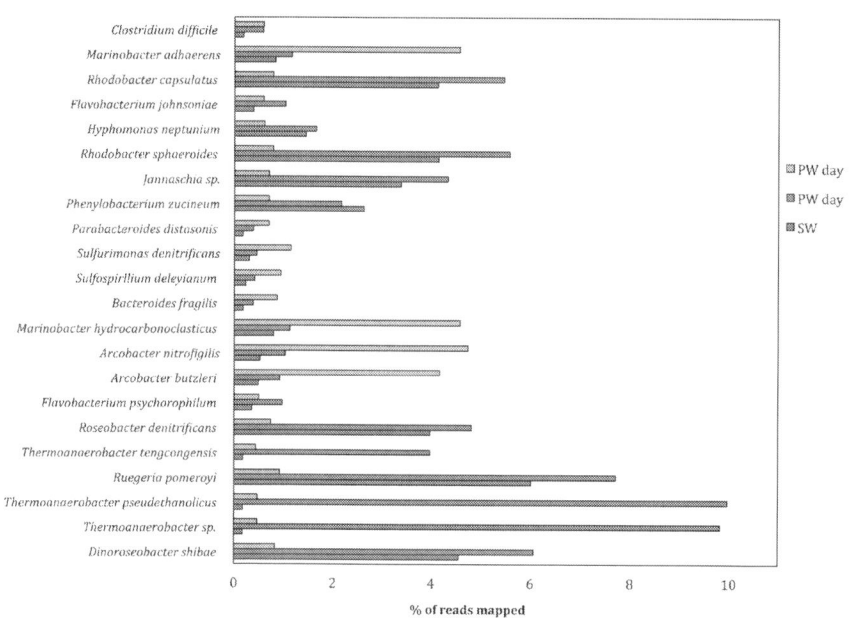

Figure 4: Read distribution for source water (SW), produced water day 1 (PW day 1) and produced water day 9 (PW day 9) samples.

Reads were mapped against reference genomes using CLC Genomic workbench version 6.5.1 using default parameters. Shown are percentages of reads mapped against each reference genome included in the analysis for all three samples.

The goal of this analysis was to provide an independent confirmation of MG-RAST results. Mapping results depend on the reference genomes selected and these reference genomes might not be the same isolates found in the environment. While reference genomes for uncultured microorganisms from oil/gas environments are limited, the positive results achieved by this mapping analysis confirm the initial taxonomic assessment.

Sulfur Metabolism Gene Mapping Results

Very few reads in all three samples were successfully mapped against the sulfur metabolism genes dsrA and dsrB. 7 reads of produced water day 1 sample and 55 reads of produced water day 9 sample were successfully mapped against the dsrA/dsrB gene of *Desulfovibrio desulfuricans* with a coverage of 28% and 78% respectively. In addition 10 reads of produced water day 9 sample were successfully mapped against the dsrA/dsrB gene of *Desulfotignum balticum* with a coverage of 19%. For aspA genes, the produced water day 9 sample showed best results with 16, 11, 9 and 6 reads successfully mapped against aspA genes of *Desulfovirbio alaskensis*, *Desulfococcus mulitvorans*, *Desulfotignum balcticum* and *Desulfobacterium autotorphicum* with a coverage of 94%, 46%, 33% and 31% respectively. Very few source water and produced water day 1 reads were mapped successfully against the aspA genes included in the analysis. These results suggest that sulfur metabolism could play a more important role in produced water day 9 sample due the higher abundance of genes associated with sulfur metabolism. Organisms that can metabolize sulfur compounds to sulfide are of interest in oil and gas environments because of their potential role in infrastructure corrosion, gas souring, worker safety as well as environmental health concerns.

Functional Classification of Metagenomes

The SEED subsystems database [42], was used to predict the metabolic potential of fracturing source water and produced water samples. Level 1 indicates the broadest set of functional categories to which sequences are assigned, and Level 2 refers to more specific functional assignments within Level 1 categories. The abundance of contigs designated to Level 1 functional categories is illustrated in Figure 5. The metabolic potential (based on Level 1 and Level 2 functional categories) between the samples was compared in a normalized manner (Figure 6, 7) to account for differences in community structure, size of the library, gene content between samples and to effectively compare low abundance functional categories [43]. Read normalization was performed within the MG-RAST analysis pipeline, in accordance with standards for metagenomic analysis.

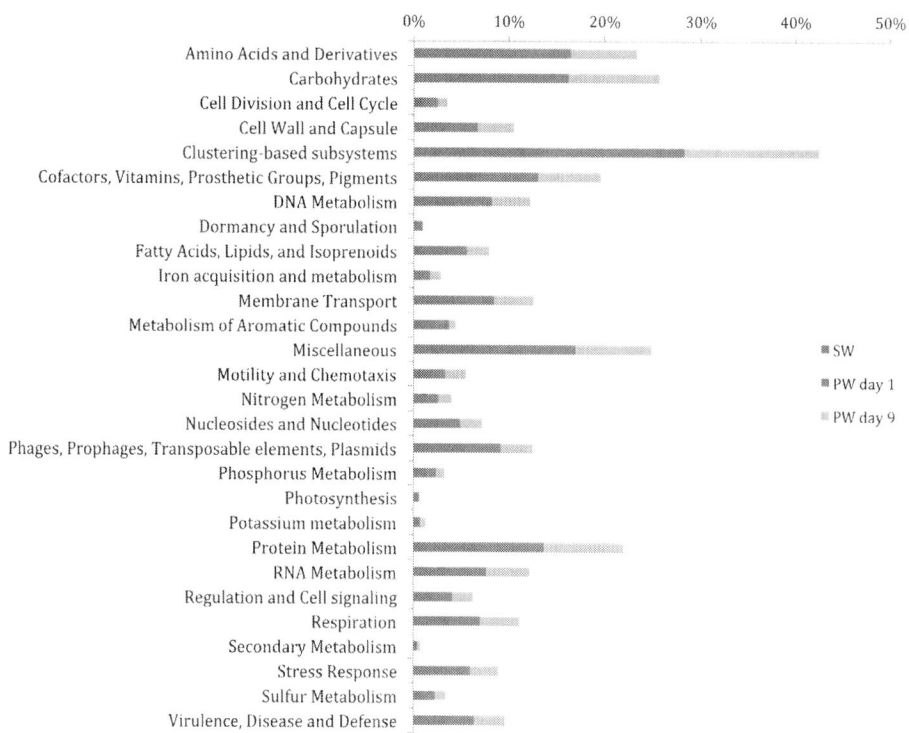

Figure 5: Actual abundance of contigs belonging to Level 1 functional categories in source water (SW), produced water day 1 (PW day 1) and produced water day 9 (PW day 9).

Functional annotations were assigned based on the Subsystems database.

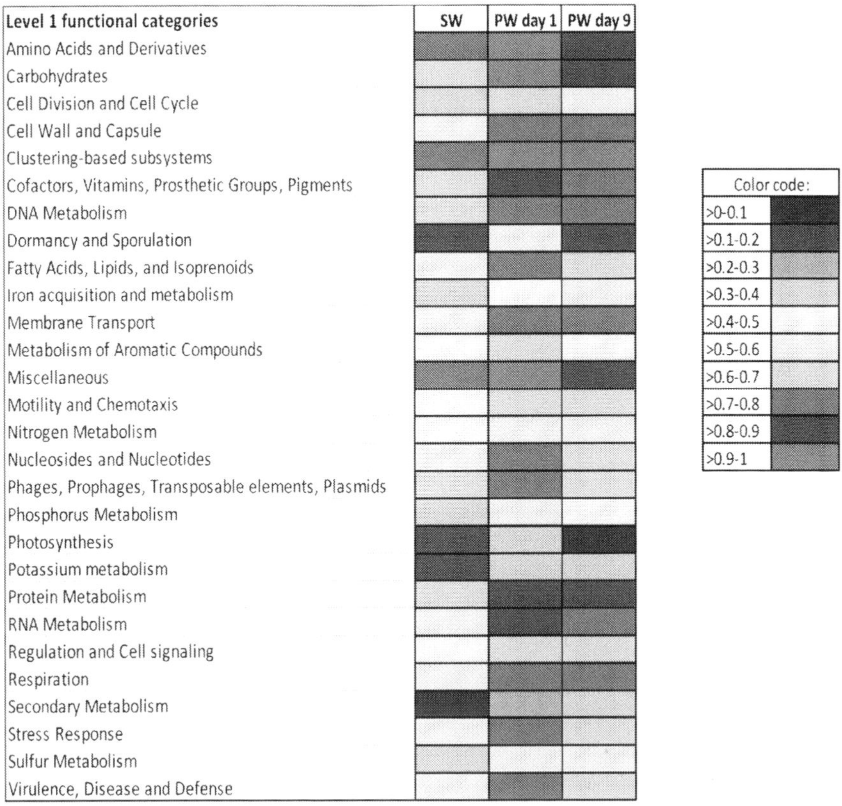

Level 1 functional categories	SW	PW day 1	PW day 9
Amino Acids and Derivatives			
Carbohydrates			
Cell Division and Cell Cycle			
Cell Wall and Capsule			
Clustering-based subsystems			
Cofactors, Vitamins, Prosthetic Groups, Pigments			
DNA Metabolism			
Dormancy and Sporulation			
Fatty Acids, Lipids, and Isoprenoids			
Iron acquisition and metabolism			
Membrane Transport			
Metabolism of Aromatic Compounds			
Miscellaneous			
Motility and Chemotaxis			
Nitrogen Metabolism			
Nucleosides and Nucleotides			
Phages, Prophages, Transposable elements, Plasmids			
Phosphorus Metabolism			
Photosynthesis			
Potassium metabolism			
Protein Metabolism			
RNA Metabolism			
Regulation and Cell signaling			
Respiration			
Secondary Metabolism			
Stress Response			
Sulfur Metabolism			
Virulence, Disease and Defense			

Color code:
- >0-0.1
- >0.1-0.2
- >0.2-0.3
- >0.3-0.4
- >0.4-0.5
- >0.5-0.6
- >0.6-0.7
- >0.7-0.8
- >0.8-0.9
- >0.9-1

Figure 6: Normalized abundance (values of 0–1) of contigs belonging to Level 1 functional categories in source water (SW), produced water day 1 (PW day 1) and produced water day 9 (PW day 9).

Functional annotations were assigned based on the Subsystems database.

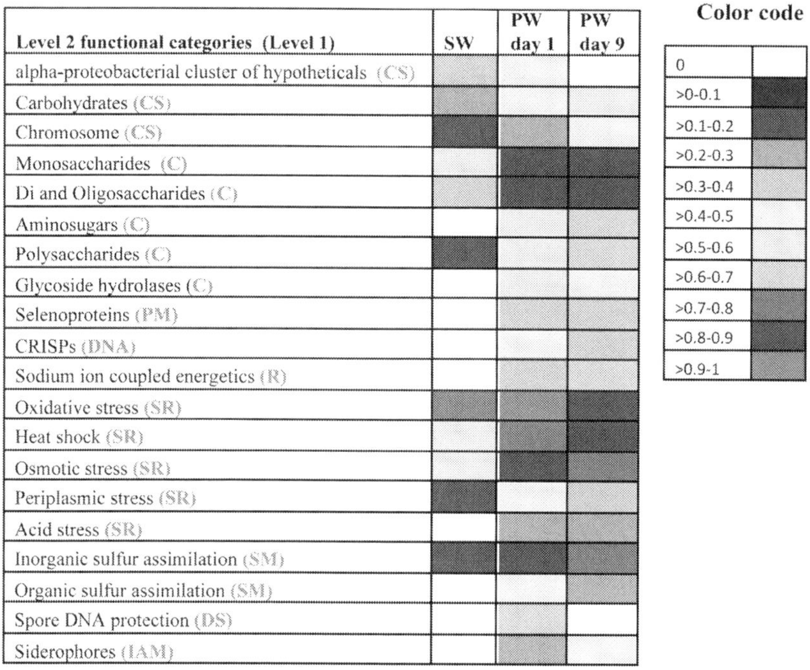

Level 2 functional categories (Level 1)	SW	PW day 1	PW day 9
alpha-proteobacterial cluster of hypotheticals (CS)			
Carbohydrates (CS)			
Chromosome (CS)			
Monosaccharides (C)			
Di and Oligosaccharides (C)			
Aminosugars (C)			
Polysaccharides (C)			
Glycoside hydrolases (C)			
Selenoproteins (PM)			
CRISPs (DNA)			
Sodium ion coupled energetics (R)			
Oxidative stress (SR)			
Heat shock (SR)			
Osmotic stress (SR)			
Periplasmic stress (SR)			
Acid stress (SR)			
Inorganic sulfur assimilation (SM)			
Organic sulfur assimilation (SM)			
Spore DNA protection (DS)			
Siderophores (IAM)			

Color code:
- 0
- >0-0.1
- >0.1-0.2
- >0.2-0.3
- >0.3-0.4
- >0.4-0.5
- >0.5-0.6
- >0.6-0.7
- >0.7-0.8
- >0.8-0.9
- >0.9-1

Figure 7: Normalized abundance (values of 0–1) of contigs belonging to selected Level 2 functional categories within associated Level 1 categories in source water (SW), produced water day 1 (PW day 1) and produced water day 9 (PW day 9).

Functional annotations were assigned based on the Subsystems database. The affiliations of Level 2 categories to Level 1 categories are coded as follows CS- Clustering based subsystems; C- Carbohydrates; PM- Protein metabolism; DNA- DNA metabolism; R- Respiration; SR- Stress response; SM- Sulfur metabolism; DS- Dormancy and sporulation; IAM- Iron acquisition and metabolism.

The five most abundant Level 1 functional categories in all three samples were found to be clustering-based subsystems (e.g. genes where functional coupling is evident but function is unknown;~14%), carbohydrate metabolism (7–9%), amino acids and derivatives (7–8%), miscellaneous (eg: genes associated with iron sulfur cluster assembly and Niacine-Choline transport and metabolism; 8–9%), protein metabolism (6–8%), suggesting the dominant role of these functional categories in all samples (Figure 5). These functional categories were similarly identified as dominant in previous studies of soil [44], [45], marine samples [24],[46], activated sludge [24], freshwater [24] and hypersaline environments [24]. Normalization of gene abundance data shows a relative increase in each of the above functional categories in the produced water samples as compared to the fracturing source water (Figure 6) implying that core systems necessary for survival are enriched in the produced water community.

While comparison of gene abundance affiliated with the dominant broad Level 1 categories suggests similar functional profiles across samples, analysis of more specific Level 2 functional categories shows sample specific differences in metabolic capabilities (Figure 7). Differences in metabolic potential indicate a selective pressure exerted in the subsurface for microbes with particular metabolic capabilities. For instance, within the Level 1 carbohydrate metabolism category, sequences related to Level 2 functional categories such as mono-, di-, oligo- and polysaccharides, and aminosugar metabolism were present in higher relative abundance in the produced water samples (Figure 7). This finding correlates well with the expected higher content of carbohydrates in produced water samples [5]. Carbohydrates and polysaccharide compounds added during hydraulic fracturing can serve as carbon and energy sources for microbial activity [8]. Within the Level 1 protein metabolism category, sequences affiliated with the Level 2 selenoprotein category were detected only in the produced water samples (Figure 7). One possible explanation is the role of selenoproteins in combating oxidative stress [47], which may arise from elevated concentrations of organic or inorganic dissolved constituents in produced water [48]. Results showed that *Rhodobacterales* were the dominant population involved in oxidative stress response in source water and produced water day 1 samples (Figure 8). However, *Alteromonadales* and *Vibrionales* were the dominant orders involved in oxidative stress response in produced water day 9 sample (Figure 8). Within the Level 1 clustering subsystem, genes affiliated with the Level 2 carbohydrate metabolism show a relative increase in the produced water samples as compared to fracturing source water (Figure 7). An increase in the relative abundance of genes related to carbohydrate metabolism in produced water compared to fracturing source water suggests the potential for utilization of hydrocarbons added either as fracturing fluid amendments or those derived from the shale formation and an overall shift to a more heterotrophic microbial community.

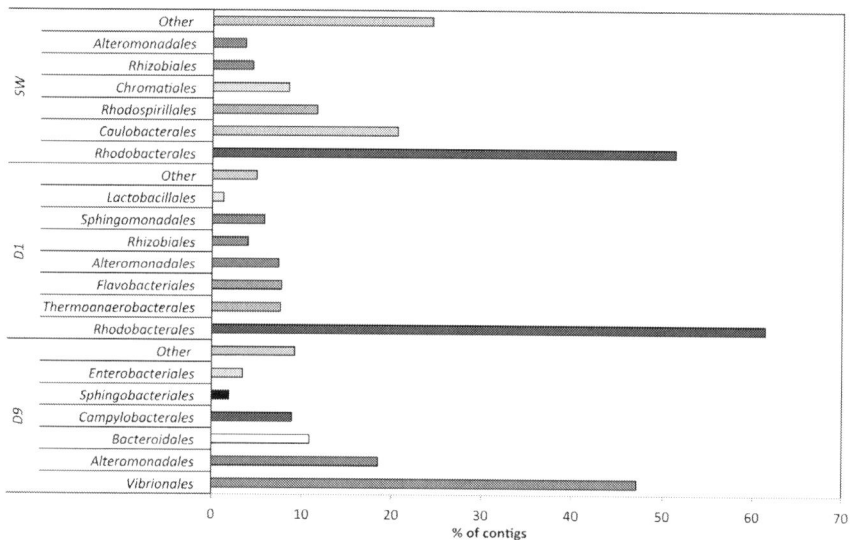

Figure 8: Taxonomic classification of oxidative stress contigs for each analyzed water sample as assigned by MGTAXA.

SW- Source water; D1- Produced water day 1; D9- Produced water day 9. Only the top six bacterial orders to which most contigs were assigned to are shown in the figure. The less abundant bacterial orders are grouped as "other".

Less abundant Level 1 functional categories showing an increase in normalized abundance in produced water samples (Figure 6) included genes affiliated with stress response (3%), respiration (3–4%), iron acquisition and metabolism (1%), sulfur metabolism (1%), and dormancy and sporulation (0.2–1%). Analysis of Level 2 functional categories within these Level 1 domains identified differences in metabolic potential between these samples (Figure 7). Within the Level 1 stress response domain, produced water samples showed a greater relative abundance of sequences affiliated with Level 2 categories such as acid stress, heat shock, periplasmic stress and osmotic stress (Figure 7). The increase in the relative abundance of these genes suggests a response to external stress experienced by the produced water microbial community. Results suggest that produced water day 1 population involved in osmotic stress response was dominated by the order *Rhodobacterales* and produced water day 9 population involved osmotic stress response was dominated by the orders *Vibrionales* and *Alteromonadales* (Figure 9). Subsurface stresses can include increased subsurface temperatures (>40°C) [49], addition of HCl and biocides to fracturing fluid, and higher concentrations of dissolved salts [5]. Within the Level 1 respiration category, sequences affiliated to the Level 2 category of sodium ion coupled energetics were undetected in fracturing source water (Na^+ 2.9 g/L) but increased in relative abundance with time in produced water samples (Na^+ concentrations in PW day 1 and day 9 were 13.9 and 43 g/L) (Figure 7). This suggests that the produced water microbial community could use sodium ion coupled energetics for their energy needs, consistent with previous observations in saline environments [50]. In the Level 1 domain of sulfur metabolism, the relative abundance of genes affiliated with Level 2 functional categories of inorganic and organic sulfur assimilation increased in produced water samples as compared to fracturing source water (Figure 7). Genes recovered from produced water day 1 show that populations involved in sulfur metabolism were dominated by the orders *Rhodobacterales* and *Thermoanaerobacterales* (Figure 10). However, sulfur metabolism in produced water day 9 samples was dominated by the orders *Vibrionales* and *Bacteroidales* (Figure 10). Within the Level 1 domain of iron metabolism, sequences affiliated with siderophores, undetected in the fracturing source water, increased with time in produced water samples (Figure 7). Siderophores are strong chelators of ferric iron secreted and are utilized by bacteria for iron metabolism [51]. Relative increase in siderophore affiliated genes correlates with an increase in total iron concentrations with time in produced water (4.2–81.6 mg/L).

Within the Level 1 dormancy and sporulation category, high relative abundance of Level 2 spore DNA protection related sequences in produced water day 1 sample (Figure 7) suggests the potential for long term dormancy of cells through DNA protection [52]. BLAT analysis [37] showed that these genes were similar to those present in *Thermoanaerobacter*, a bacterial order that constituted 16% of the total community in this sample (Figure 2). An increase in the relative abundance of spore forming bacteria and genes affiliated with sporulation and dormancy is an important consideration in biocide application, and may provide an explanation for the previously observed limited efficacy of biocides [5].

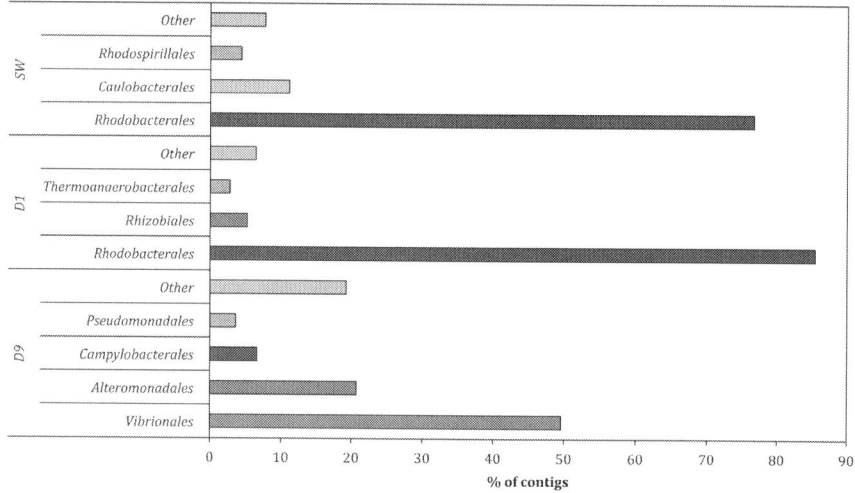

Figure 9: Taxonomic classification of osmotic stress contigs for each analyzed water sample as assigned by MGTAXA.

SW- Source water; D1- Produced water day 1; D9- Produced water day 9. Only the top four bacterial orders to which most contigs were assigned to are shown in the figure. The less abundant bacterial orders are grouped as "other".

The Functional Potential of Microbial Communities in ... 245

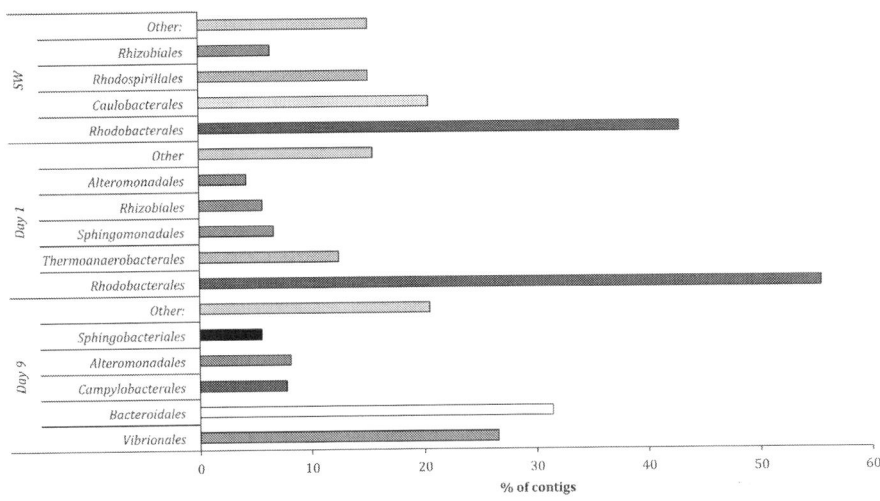

Figure 10: Taxonomic classification of sulfur metabolism contigs for each analyzed water sample as assigned by MGTAXA.

SW- Source water; D1- Produced water day 1; D9- Produced water day 9. Only the top five bacterial orders to which most contigs were assigned to are shown in the figure. The less abundant bacterial orders are grouped as "other".

Concluding Remarks.

This study is the first shotgun metagenomic analysis of produced water from hydraulic fracturing for natural gas production and provides novel insights on taxonomic and functional potential of this pertinent yet unexplored environment. Taxonomic analysis showed that *Bacteria* constituted the dominant (>98%) domain in both fracturing source water and produced water samples. Results demonstrated the emergence of distinct bacterial classes and orders in the produced water samples and fracturing source water samples. These bacterial taxa were consistent with results from a previous 16S rRNA gene based survey of these samples [5]. The metabolic profile showed both a relative increase and functional changes in genes responsible for carbohydrate metabolism, respiration, sporulation and dormancy, iron acquisition and metabolism, stress response and sulfur metabolism in the produced water samples as compared to the fracturing source water sample. These results suggest that the microbial community is responsive to changes in hydrocarbon content, induced stresses such as increase in temperature, addition of biocides, and an increase in concentration of dissolved salts such as iron and sulfur. The detection of genes affiliated with sodium ion coupled energetics exclusively in the produced water samples suggests the use of sodium ion based energetics by microorganisms in these sodium rich environments. Understanding the evolving metabolic capabilities of microbial communities in produced water will help the industry and its regulators improve environmental and economic sustainability of oil and gas extraction through more informed water management decisions.

AUTHOR CONTRIBUTIONS

Conceived and designed the experiments: AMM KBG KJB RWH. Performed the experiments: AMM. Analyzed the data: AMM KJB DL. Contributed reagents/materials/analysis tools: KBG KJB. Wrote the paper: AMM KJB DL RWH KBG.

REFERENCES

1. Veil JA (2010), Water Management Technologies Used by Marcellus Shale Gas Producers, ANL/EVS/R-10/3, prepared by Environmental Science Division, Argonne National Laboratory for the U.S. Department of Energy, Office of Fossil Energy, National Energy Technology Laboratory, July ANL/EVS/R-10/3.

2. Arthur JD, Bohm B, Coughlin B, Layne M, Cornue D (2009) Evaluating the Environmental Implications of Hydraulic Fracturing in Shale Gas Reservoirs, *SPE 121038*, In: SPE Americas Environmental and Safety Conference. San Antonio, TX, March 23–25.

3. Gregory KB, Vidic RD, Dzombak DA (2011) Water Management Challenges Associated with the Production of Shale Gas by Hydraulic Fracturing. Elements 7: 181–186. doi: 10.2113/gselements.7.3.181
4. Barbot E, Vidic N, Gregory KB, Vidic RD (2013) Spatial and Temporal Correlation of Water Quality Parameters of Produced Waters from Devonian-Age Shale following Hydraulic Fracturing. Environ Sci Technol 47: 2562–2569. doi: 10.1021/es304638h
5. Murali Mohan A, Hartsock A, Bibby K, Hammack RW, Vidic RD, et al. (2013) Microbial Community Changes in Hydraulic Fracturing Fluids and Produced Water from Shale Gas Extraction. Environ Sci Technol 47(22): 13141–13150. doi: 10.1021/es402928b
6. Soeder DJ, Kappel WM (2009) USGS Fact Sheet 2009–3032.
7. Hill D, Lombardi T, Martin J (2004) Fractured Shale Gas Potential In New York. Northeastern Geol. Environ Sci 26: 57–78.
8. Moore SL, Cripps CM (2010) Bacterial Survival in Fractured Shale Gas Wells of the Horn River Basin (CSUG/SPE 137010). CSUG 1–14. doi: 10.2118/137010-pa
9. Murali Mohan A, Hartsock A, Hammack RW, Vidic RD, Gregory KB (2013) Microbial Communities in Flowback Water Impoundments from Hydraulic Fracturing for Recovery of Shale Gas. FEMS Microbiology Ecology 86(3): 567–580. doi: 10.1111/1574-6941.12183
10. Kermani M, Harrop D (1996) The impact of corrosion on oil and gas industry. SPE Production Facilities 11: 186–190. doi: 10.2118/29784-pa
11. Little BJ, Lee JS (2007) Microbiologically influenced corrosion. Wiley and Sons Inc., Hoboken, NJ.
12. Fichter JK, Johnson K, French K, Oden R (2008) Use of Microbiocides in Barnett Shale Gas Well Fracturing Fluids to Control Bacteria Related Problems (Paper No. 08658). In NACE International Corrosion Conference and Expo pp. 1–14.

13. Roberge PR (2000) Handbook of Corrosion Engineering. McGraw-Hill, New York.
14. Struchtemeyer CG, Morrison MD, Elshahed MS (2012) A critical assessment of the efficacy of biocides used during the hydraulic fracturing process in shale natural gas wells. International Biodeterioration & Biodegradation 71: 15–21. doi: 10.1016/j.ibiod.2012.01.013
15. Struchtemeyer CG, Elshahed MS (2012) Bacterial communities associated with hydraulic fracturing fluids in thermogenic natural gas wells in North Central Texas, USA. FEMS Microbiology Ecology 81: 13–25. doi: 10.1111/j.1574-6941.2011.01196.x
16. Williams TM, Mcginley HR (2010) Deactivation of Industrial Water Treatment Biocides (Paper No. 10049). In NACE International Corrosion Conference and Expo pp. 1–15.
17. Davis JP, Struchtemeyer CG, Elshahed MS (2012) Bacterial communities associated with production facilities of two newly drilled thermogenic natural gas wells in the Barnett Shale (Texas, USA). Microbial Ecology 64: 942–954. doi: 10.1007/s00248-012-0073-3
18. Dahle H, Garshol F, Madsen M, Birkeland NK (2008) Microbial community structure analysis of produced water from a high-temperature North Sea oil-field. Antonie van Leeuwenhoek 93: 37–49. doi: 10.1007/s10482-007-9177-z
19. Pham VD, Hnatow LL, Zhang S, Fallon RD, Jackson SC, et al. (2009) Characterizing microbial diversity in production water from an Alaskan mesothermic petroleum reservoir with two independent molecular methods. Environmental Microbiology 11: 176–187. doi: 10.1111/j.1462-2920.2008.01751.x
20. Grabowski A, Nercessian O, Fayolle F, Blanchet D, Jeanthon C (2005) Microbial diversity in production waters of a low-temperature biodegraded oil reservoir. FEMS Microbiology Ecology 54: 427–443. doi: 10.1016/j.femsec.2005.05.007
21. van der Kraan GM, Bruining J, van Loosdrecht MCM, Muyzer

G (2010) Microbial diversity ofan oil water processing site and its associated oil field: the possible role of microorganisms as information carriers from oil-associated environments. FEMS Microbiology Ecology 71: 428–443. doi: 10.1111/j.1574-6941.2009.00813.x

22. Gittel A, Sørensen KB, Skovhus TL, Ingvorsen K, Schramm A (2009) Prokaryotic community structure and sulfate reducer activity in water from high-temperature oil reservoirs with and without nitrate treatment. Appl Environ Microbiol 75: 7086–7096. doi: 10.1128/aem.01123-09

23. Hendrickson H (2009) Order and disorder during *Escherichia coli* divergence. PLoS genetics 5: e1000335 Available: http://www.plosgenetics.org/article/info%3Adoi%2F10.1371%2Fjournal.pgen.1000335.

24. Dinsdale EA, Edwards RA, Hall D, Angly F, Breitbart M, et al. (2008) Functional metagenomic profiling of nine biomes. Nature 452: 629–632. doi: 10.1038/nature06810

25. DeLong EF, Preston CM, Mincer T, Rich V, Hallam SJ, et al. (2006) Community genomics among stratified microbial assemblages in the ocean's interior. Science 311: 496–503. doi: 10.1126/science.1120250

26. Tringe SG, Rubin EM (2005) Metagenomics: DNA sequencing of environmental samples. Nature reviews. Genetics 6: 805–814. doi: 10.1038/nrg1709

27. Wegley L, Edwards R, Beltran Rodriguez-Brito1 HL, Rohwer F (2007) Metagenomic analysis of the microbial community associated with the coral *Porites astreoides*. Environmental Microbiology 9: 2707–2719. doi: 10.1111/j.1462-2920.2007.01383.x

28. Yu K, Zhang T (2012) Metagenomic and metatranscriptomic analysis of microbial community structure and gene expression of activated sludge. PloS one 7: e38183 Available: http://www.plosone.org/article/info%3Adoi%2F10.1371%2Fjournal.pone.0038183.

29. Yergeau E, Hogues H, Whyte LG, Greer CW (2010) The functional potential of high Arctic permafrost revealed by metagenomic sequencing, qPCR and microarray analyses. The ISME journal 4: 1206–1214. doi: 10.1038/ismej.2010.41
30. Dongshan AN, Caffrey SM, Soh J, Agrawal A, Brown D, et al. (2013) Metagenomics of Hydrocarbon Resource Environments Indicates Aerobic Taxa and Genes to be Unexpectedly Common. Environ Sci Technol 47: 10708–10717. doi: 10.1021/es4020184
31. Dong Y, Kumar CG, Chia N, Kim PJ, Miller P, et al. (2013) *Halomonas sulfidaeris*-dominated microbial community inhabits a 1.8 km-deep subsurface Cambrian Sandstone reservoir. Environmental Microbiology 16(6): 1695–1708. doi: 10.1111/1462-2920.12325
32. Lamendella R, Domingo JWS, Ghosh S, Martinson J, Oerther DB (2011) Comparative fecal metagenomics unveils unique functional capacity of the swine gut. BMC Microbiology 11: 103 Available: http://www.biomedcentral.com/1471-2180/11/103.
33. Schmieder R, Lim YW, Rohwer F, Edwards R (2010) TagCleaner: Identification and removal of tag sequences from genomic and metagenomic datasets. BMC Bioinformatics 11: 341 Available: http://www.ncbi.nlm.nih.gov/pmc/articles/PMC2910026/.
34. Goecks J, Nekrutenko A, Taylor J (2010) The Galaxy Team (2010) Galaxy: A comprehensive approach for supporting accessible, reproducible, and transparent computational research in the life sciences. Genome biology 11: R86 Available:http://genomebiology.com/2010/11/8/R86.
35. Zerbino DR (2010) Using the Velvet de novo assembler for short-read sequencing technologies. Curr Protoc Bioinformatics 31: 11.5.1–11.5.12 Available:http://www.ncbi.nlm.nih.gov/pmc/articles/PMC2952100/.
36. Meyer F, Paarmann D, D'Souza M, Olson R, Glass EM, et al. (2008) The metagenomics RAST server - a public

37. Kent WJ (2002) BLAT—The BLAST-Like Alignment Tool. Genome Research 12: 656–664 Available: http://www.ncbi.nlm.nih.gov/pmc/articles/PMC187518/.
38. Tovchigrechko A, Sul SJ, MGTAXA- A free software for taxonomic classification of metagenomic sequences with machine learning techniques. Available:http://andreyto.github.io/mgtaxa/
39. Brady A, Salzberg SL (2009) Classification with interpolated markov models. Nature Methods 6(9): 673–676. doi: 10.1038/nmeth.1358
40. Giardine B, Riemer C, Hardison C, Burhans R, Elnitski L (2005) Galaxy: A platform for interactive large scale genome analysis. Genome Research 15: 1451–1455 Available:http://www.ncbi.nlm.nih.gov/pmc/articles/PMC1240089/.
41. CLC Genomics Workbench, "Version 6.5.1", CLC bio A/S Science Park Aarhus Finlandsgade, 10–12. Available: http://www.clcbio.com/products/clc-genomics-workbench/
42. Overbeek R, Begley T, Butler RM, Choudhuri JV, Chuang HY, et al. (2005) The subsystems approach to genome annotation and its use in the project to annotate 1000 genomes. Nucleic acids research 33: 5691–5702. doi: 10.1093/nar/gki866
43. Shi Y, Tyson GW, Eppley JM, DeLong EF (2011) Integrated metatranscriptomic and metagenomic analyses of stratified microbial assemblages in the open ocean. The ISME journal 5: 999–1013. doi: 10.1038/ismej.2010.189
44. Delmont TO, Prestat E, Keegan KP, Faubladier M, Robe P, et al. (2012) Structure, fluctuation and magnitude of a natural grassland soil metagenome. The ISME Journal 6(9): 1677–1687. doi: 10.1038/ismej.2011.197
45. Urich T, Lanzén A, Qi J, Huson DH, Schleper C, et al. (2008) Simultaneous assessment of soil microbial community structure and function through analysis of the meta-transcriptome.

(continued from previous page) resource for the automatic phylogenetic and functional analysis of metagenomes. *BMC Bioinformatics* 2008 9: 386 Available:http://www.biomedcentral.com/1471-2105/9/386.

PloS one 3: e2527 Available:http://www.plosone.org/article/info%3Adoi%2F10.1371%2Fjournal.pone.0002527.

46. Gilbert JA, Field D, Huang Y, Edwards R, Li W, et al. (2008) Detection of large numbers of novel sequences in the metatranscriptomes of complex marine microbial communities. PloS one 3: e3042 Available: http://www.plosone.org/article/info%3Adoi%2F10.1371%2Fjournal.pone.0003042.

47. Lu J, Holmgren A (2009) Selenoproteins. The Journal of biological chemistry 284: 723–727. doi: 10.1074/jbc.r800045200

48. Valavanidis A, Vlahogianni T, Dassenakis M, Scoullos M (2006) Molecular biomarkers of oxidative stress in aquatic organisms in relation to toxic environmental pollutants. Ecotoxicol Environ Saf 64: 178–189. doi: 10.1016/j.ecoenv.2005.03.013

49. Driscoll FG (1986) Groundwater and Wells. Johnson Filtration Inc.: St Paul, MN.

50. Kogure K (1998) Bioenergetics of marine bacteria. Current Opinion in Biotechnology 9: 278–282. doi: 10.1016/s0958-1669(98)80059-1

51. Sandy M, Butler A (2010) Microbial Iron Acquisition: Marine and Terrestrial Siderophores. Chem Rev 109: 4580–4595. doi: 10.1021/cr9002787

52. Setlow P (1992) Mini Review: I Will Survive: Protecting and Repairing Spore DNA. Journal of Bacteriology 174: 2737–2741.

Citations

CHAPTER 1

Amir Shojaei, Arash Dahi Taleghani, Guoqiang Li, A continuum damage failure model for hydraulic fracturing of porous rocks, International Journal of Plasticity, Volume 59, August 2014, Pages 199-212, ISSN 0749-6419, http://dx.doi.org/10.1016/j.ijplas.2014.03.003.

CHAPTER 2

Md. Mofazzal Hossain, M.K. Rahman, Numerical simulation of complex fracture growth during tight reservoir stimulation by hydraulic fracturing, Journal of Petroleum Science and Engineering, Volume 60, Issue 2, February 2008, Pages 86-104, ISSN 0920-4105, http://dx.doi.org/10.1016/j.petrol.2007.05.007.

CHAPTER 3

Xiaodan Ma, Eduardo Gildin, Tatyana Plaksina, Efficient optimization framework for integrated placement of horizontal wells and hydraulic fracture stages in unconventional gas reservoirs, Journal of Unconventional Oil and Gas Resources, Volume 9, March 2015, Pages 1-17, ISSN 2213-3976, http://dx.doi.org/10.1016/j.juogr.2014.09.001.

CHAPTER 4

Hammond, P. and Field, M. (2014) A Reinterpretation of Historic Aquifer Tests of Two Hydraulically Fractured Wells by Application of Inverse Analysis, Derivative Analysis, and Diagnostic Plots. Journal of Water Resource and Protection, 6, 481-506. doi: 10.4236/jwarp.2014.65048.

CHAPTER 5

D. Kwak, M. Kim, J. Kim, Y. Oh, S. Noh, B. So, S. Jung, S. Jung and S. Chae, "Evaluation of Methane Yield on Mesophilic-Dry Anaerobic Digestion of Piggery Manure Mixed with Chaff for Agricultural Area," Advances in Chemical Engineering and Science, Vol. 3 No. 4, 2013, pp. 227-235. doi: 10.4236/aces.2013.34029.

CHAPTER 6

M. Hameed and M. Jawad, "Hydrodynamics of Liquid Film in Helical Tubes," Advances in Chemical Engineering and Science, Vol. 2 No. 1, 2012, pp. 74-81. doi: 10.4236/aces.2012.21009.

CHAPTER 7

Mohan AM, Bibby KJ, Lipus D, Hammack RW, Gregory KB (2014) The Functional Potential of Microbial Communities in Hydraulic Fracturing Source Water and Produced Water from Natural Gas Extraction Characterized by Metagenomic Sequencing. PLoS ONE 9(10): e107682. doi:10.1371/journal.pone.0107682.

CHAPTER 8

N. Hadia, L. Chaudhari, Sushanta K. Mitra, M. Vin

CHAPTER 9

Yu-Chao Zeng, Neng-You Wu, Zheng Su, Jian Hu, Numeri

Index

A
Aquifer test methods 122

B
Boundary element method (BEM) 45, 49

C
Coil diameter 203, 218
Continuum damage mechanics (CDM) 1, 3
Covariance matrix adaptation evolution strategy (CMA-ES) 82, 84

E
Environmental pollution 174
Exponential Model (EM) 187

F
Fluid flow 202

G
Genetic Algorithm (GA) 82, 84, 90

H
Hydraulic fracture (HF) 82

I

Infinite acting radial flow (IARF) 124

L

Linear elastic fracture mechanics (LEFM) 44
Local grid refinement feature (LGR) 99

M

Massive sediment 159
Modified Gompertz Model (MGM) 187

N

Natural fracture 43
Natural gas 226, 227, 228, 235, 245, 248

Net present value (NPV) 82, 83

P

Petroleum production 42

S

Simultaneous perturbation stochastic approximation (SPSA) 82, 88
Single, horizontal fracture model (SHF) 141
Source water (SW) 230, 232, 235, 236, 238, 239, 240
Stimulated reservoir volume (SRV) 99
Sum of the squared residuals (RSS) 129

V

Vertical tube 201
Volatile solid (VS) 174, 193

W

Wastewater 226